J. Bullerdiek S. Bartnitzke (Eds.)

Chromosome 12 Aberrations in Human Solid Tumors

Cytogenetics and Molecular Genetics

With a Foreword by H. Van den Berghe

With 68 Figures, Some in Colour and 12 Tables

Springer-Verlag Berlin Heidelberg GmbH

Prof. Dr. J. BULLERDIEK
Dr. S. BARTNITZKE
Universität Bremen
Zentrum f. Humangenetik u. Genetische Beratung
Leobener Straße
28359 Bremen, Germany

ISBN 978-3-662-06257-9 ISBN 978-3-662-06255-5 (eBook)
DOI 10.1007/978-3-662-06255-5

Originally published by Springer-Verlag Berlin Heidelberg New York in 1994
Softcover reprint of the hardcover 1st edition 1994

Typesetting: Best-set Typesetter Ltd., Hong Kong
SPIN: 10066684 19/3130/SPS - 5 4 3 2 1 0 - Printed on acid-free paper

Foreword

For cancer genetics, chromosome 12 is an extraordinarily interesting chromosome. Both the short arm and the long arm obviously carry a series of sequences which play a major role in the pathogenetic mechanisms of a large variety of malignancies.

On the short arm there is at least one so-called lymphoid and one myeloid breakpoint relevant for hematologic malignancies. The latter breakpoint may be involved particularly in iatrogenic or secondary neoplasias. With regard to solid tumors, the iso(12p) is specific for germ cell tumors in both sexes, be they gonadal or extragonadal.

Much more information, however, is available for the long arm of the chromosome, particularly the region 12q13–15. From the available data two malignant mesenchymal tumors, myxoid liposarcoma and the clear-cell sarcoma, plus a series of benign tumors seem to cluster in this region. Among the benign tumors are a group of lipomas, a group of uterine leiomyomas, a group of pleomorphic adenomas of the salivary gland, and perhaps some more tumors. The same region appears to be involved in well-differentiated liposarcoma.

Trisomy of the whole chromosome characterizes a group of tumors of the female urogenital tract, as well as some CLL. Finally an intermediate filament associated protein expressed in Reed-Sternberg cells maps to 12q24 which may open new ways to investigate the pathogenesis of Hodgkin lymphoma.

H. Van den Berghe

Preface

Chromosome 12 changes, including trisomies, isochromosomes of the short arm, or translocations affecting 12q13–15, are frequent findings in human solid tumors of both epithelial and mesenchymal origin. An outstanding example of the latter type is the translocation t(12;14)(q14–15;q23–24) in uterine leiomyoma, which is considered to be the most frequent, specific chromosomal translocation in human tumors at all.

It is the aim of this book to provide an introduction to this interesting area of tumor genetics. At first glance the theme has two main aspects: the field of microscopically visible chromosomal changes and the field of "pure" molecular genetics. The book has three major parts however: In the first part, "Cytogenetic Background," the cytogenetics of chromosome 12 changes in human solid tumors are introduced. The contributions in this chapter all deal mainly with findings which can be seen at the microscopic level. In the second and the third parts, "The Molecular Genetics of Chromosome 12" and "Gene Analysis: Methods and Facts," attempts to understand the molecular alterations underlying the visible chromosomal alterations are described. Although mainly focused on the results of these studies, we felt that both parts should also include a description of methodological approaches as, for example, establishing cell lines or chromosomal microdissection to show the pathways by which cytogenetic findings can be traced directly back to molecular genetic terms. We hope that what appears at first glance to be independent areas will, in the end, turn out to be like the major branches of a thick trunk, the trunk of tumor cytogenetics. We feel that almost all the contributions in this book will also reveal that tumor cytogenetics is not an old fashioned way of studying the genetic background of tumor cells, which is now rapidly being replaced by molecular genetics. If that were true, neither chromosomal microdissection nor fluorescence in situ hybridization would exist. Who could microdissect a chromosome without recognizing it?

However, though inevitably incomplete, we hope that this book gives an informative overview, summarizing the field of chromosome 12 changes in human solid tumors. To keep the reader informed about recent papers which have appeared in this field or aspects not covered in the major contributions, a selected further reading section has been added at the end of the book. Of course, this is also inevitably incomplete, which by no means indicates that those papers which were not included are less important.

Without the help of many members of our laboratory this book would hardly exist. First of all, we wish to express our special gratitude to Dr. B. Kazmierczak, whose help has been invaluable. We would also especially like to mention Dr. B. Rommel and Ms. N. Reimann for critically reading parts of the manuscript and Dr. G. Belge and V. Rippe for suggestions and criticism. Our sincere thanks are due to our secretaries, Ms. A. Schneider-Uhlhorn, Ms. L. Tanner, and Ms. B. Kreihe, who typed part of the manuscript. The workshop "Cytogenetics and Molecular Genetics of Chromosome 12 Changes in Human Solid Tumors" was supported by the European Community through the 4th Medical and Health Research Programme, the University of Bremen, and the Förderkreis Humangenetik Bremen EV.

Bremen, 1994

Sabine Bartnitzke
Jörn Bullerdiek

Contents

Cytogenetic Background

List of Contributors*

* The addresses of the authors are given on the first page of each contribution.
[1] Page on which contribution begins.

Cytogenetic Background

Specificity and Implications of Chromosome 12 Abnormalities in Pleomorphic Adenomas

G. Stenman[1], J. Bullerdiek[2], S. Bartnitzke[2], P. Sahlin[1], E. Röijer[1], and J. Mark[3]

1 Clinicopathological Characteristics of Pleomorphic Adenomas

The pleomorphic adenoma, or mixed salivary gland tumor, is a benign epithelial tumor originating from the major and minor salivary glands. Eighty-five percent of these tumors are found in the parotid gland, 10% in the minor salivary glands, and 5% in the submandibular gland (Seifert et al. 1986). The pleomorphic adenoma is the most common type of salivary gland tumor, accounting for almost 50% of all neoplasms in these organs (Seifert et al. 1986). Pleomorphic adenomas occur most frequently in women in the fourth decade of life, but they can also be found in children and elderly persons of either sex. The tumor is almost exclusively benign; malignancies are rare.

Clinically, pleomorphic adenomas present as a solitary, slowly growing painless mass. Grossly, they are encapsulated and usually well circumscribed, with yellow-to-white cut solid surfaces. Microscopically, adenomas are composed of epithelial and myoepithelial cells arranged in a variety of patterns in a matrix of mucoid, myxoid, chondroid and, on rare occasions, even osteoid tissue. The proportion of the various cell elements varies considerably from one tumor to another.

Pleomorphic adenomas display immunoreactivity for a number of epithelial-glandular markers such as keratin, carcinoembryonic antigens (CEA), tissue polypeptide antigen (TPA), epithelial membrane antigen (EMA), lactoferrin, lectin receptors, and secretory component (see references in Seifert et al. 1986). Immunohistochemical, cytochemical, and ultrastructural studies have shown that the major cell components in pleomorphic adenomas are ductal and myoepithelial cells (Seifert et al. 1986). These cells are thought to originate from multipotent cells in the terminal duct system and are likely to be responsible for the marked histological diversity observed. Although the etiology of pleomorphic adenomas is un-

[1] Laboratory of Cancer Genetics, Department of Pathology, Sahlgrenska University Hospital, S-41345 Gothenburg, Sweden
[2] Center of Human Genetics and Genetic Counselling, University of Bremen, Bremen, Germany
[3] Department of Pathology, Central Hospital, Skövde, Sweden

known, it is worth noting that this particular type of neoplasm is extremely rare in animals.

2 Primary Chromosomal Abnormalities

The first type of benign tumor in which clonal chromosome abnormalities were detected was the meningioma. Banding analysis of large series of meningiomas from two independent laboratories clearly showed a frequent loss of one chromosome 22 (Mark et al. 1972; Zankl and Zang 1972). Despite this observation, most cytogeneticists continued to believe benign tumors to be by definition chromosomally normal. It was not until convincing data on a second type of benign tumor, the pleomorphic adenoma, was presented (Mark et al. 1980; Stenman et al. 1984) that it became generally accepted that other benign tumor types might also show nonrandom chromosome changes. Recently, extensive data on several other types of benign tumors (Mark et al. 1990; Nilbert and Heim 1990; Sreekantaiah et al. 1991) have unequivocally confirmed the observations made previously on meningiomas and pleomorphic adenomas. Moreover, it has appeared that as a rule the changes observed in benign tumors are more specific than those observed in most malignant tumors.

Since the original report on chromosome changes in pleomorphic adenomas appeared in 1980, about 470 adenomas have been analyzed cytogenetically in two independent laboratories in Sweden and Germany (Sandros et al. 1990; Mark et al., unpublished data; Bullerdiek et al. 1993). These studies have revealed a highly specific pattern of rearrangements involving three particular chromosome regions, i.e., 3p21, 8q12 and 12q13–15. The deviations are almost invariably of a structural character; approximately 90% are reciprocal translocations.

In addition to a group of adenomas with apparently normal diploid stemlines (20–50% of the cases in different series), there are at least three major subgroups of cytogenetically abnormal adenomas: (1) a large group of tumors with chromosome 8 abnormalities, usually consisting of translocations of the 8q12→qter segment. The most common abnormality in this group is a t(3;8)(p21;q12); (2) tumors with chromosome 12 abnormalities, usually consisting of translocations of the 12q13–15→qter segment; (3) tumors with nonrecurrent clonal abnormalities. Secondary chromosomal rearrangements, sometimes including dicentrics, ring chromosomes, and dmin chromosomes, have been found in about one-third of the tumors. The regularity and specificity of the recurrent 3p21, 8q12 and 12q13–15 rearrangements in pleomorphic adenomas strongly indicate that these chromosome regions carry genes of crucial importance for the evolution of these tumors. We will consider the specificity and implications of chromosome 12 abnormalities and the molecular interpretations of these rearrangements below.

3 Chromosome 12 Abnormalities

The pleomorphic adenoma was the first solid tumor reported to display consistent chromosome 12q13–15 abnormalities (Mark et al. 1980; Stenman et al. 1984; Bullerdiek et al. 1987). Today we know that this region appears to be a hot spot for chromosomal recombination in several other neoplastic conditions, including lipoma, leiomyoma, myxoid liposarcoma, and clear-cell sarcoma (Turc-Carel et al. 1986; Sreekantaiah et al. 1991; Nilbert and Heim 1990; Stenman et al. 1993). Swedish and German research provide data on a total of 55 pleomorphic adenomas with chromosome 12 abnormalities; 33 adenomas observed in Germany and 22 in Sweden (Sandros et al. 1990; Mark et al., unpublished data; Bullerdiek et al. 1993 and unpublished data). Chromosome 12 changes were observed in 12% of the German material; the frequency reported by Swedish research was 11%. So far, only structural rearrangements have been detected.

The distribution of breakpoints on chromosome 12 is clearly nonrandom and demonstrates that 75% of the breaks have occurred in the 12q13–15 region (Fig. 1). There are also two minor clusters of breakpoints at 12p13 and 12q24, representing 9% and 6% of the breaks respectively. The breakpoints in the 12q13–15 region are variable, although most breaks seem to occur in the distal part at 12q14–15. The reason for this variation is obscure

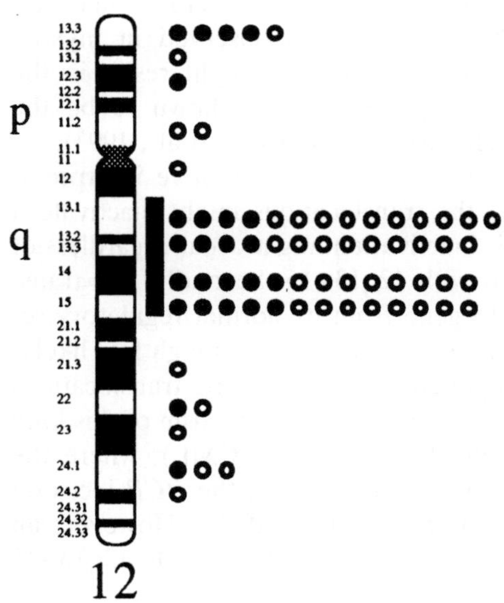

Fig. 1. Distribution of breakpoints on chromosome 12 in 55 pleomorphic adenomas. The *vertical line* to the *right* of the chromosome indicates that the breakpoints have been mapped to the segment 12q13–15 rather than to an individual band. Swedish cases are indicated by *filled circles* and German cases by *open circles*

– either there is a true variation in the location of the breakpoints, or the breakpoints are in fact the same. This is impossible to establish definitively because of the limited resolution of the banding pattern in this region. Alternatively, the breakpoints are identical; due to subsequent intrachromosomal rearrangements, e.g., deletions, inversions, or duplications, however, they seem to vary. Recent studies using fluorescence in situ hybridization (FISH) of the gene order within 12q13–15 suggest that this might be the case in certain lipomas with abnormalities of 12q13–15 (Pedeutour et al. 1992). The idea of a common breakpoint at the cytogenetic level is in keeping with current knowledge of translocations occurring, e.g., in chronic myeloid leukemia, Burkitt's lymphoma, Ewing's sarcoma, and myxoid liposarcoma (Dalla-Favera et al. 1987; Groffen et al. 1989; Zucman et al. 1992; Åman et al. 1992).

Translocations with breakpoints at 12q13–15 have been detected in 34 adenomas (Figs. 2,3). In 26 cases these were the sole anomalies, indicating that they represent primary cytogenetic events. Five different translocations were recurrent: (a) t(9;12)(p12–22;q13–15) [five cases, including a complex t(X;12;9;10)]; (b) t(1;12)(p13–22;q13–15) [three cases]; (c) t(5;12)(q13–15;q13–14) [three cases, including a complex t(1;12;5)]; (d) t(12;13)(q14–15;q32) [two cases]; (e) t(12;14)(q15;q21–23) [two cases]. In addition to chromosomes 1, 5, 9, 13, and 14, there are 12 other chromosomes that have acted as translocation partners with 12q13–15→qter (Fig. 4).

There seems to be a profound difference in this respect between benign and malignant tumors with 12q13–15 abnormalities, inasmuch as the translocation partners in myxoid liposarcoma (chromosome 16) and clear-cell sarcoma (chromosome 22) are almost invariably consistent, while the translocation partners in pleomorphic adenoma, lipoma, and leiomyoma vary considerably. The consistency of the translocations observed in myxoid liposarcoma and clear-cell sarcoma suggests that they might result in the formation of a fusion gene. Indeed, this was recently shown to be the case in myxoid liposarcoma with t(12;16)(q13;p11) (Crozat et al., 1993).

In the case of benign tumors, there are several alternative interpretations that can be considered. Firstly, the translocations might inactivate a suppressor gene at 12q13–15, and this inactivation might occur regardless of the chromosome that has recombined with 12q13–15. Secondly, the translocations might separate the 12q13–15 gene from its normal regulatory sequences, leading to aberrant gene expression. A third, although less likely, explanation is that, despite their cytogenetic differences, the translocations at the molecular level are identical or very similar. A mechanism of this kind has been observed in, e.g., chronic myeloid leukemia (CML), where the variant translocations also show translocation of *ABL* to the *BCR* locus on chromosome 22 at the molecular level (Dubé et al. 1989). However, an important difference between CML and benign tumors with 12q13–15

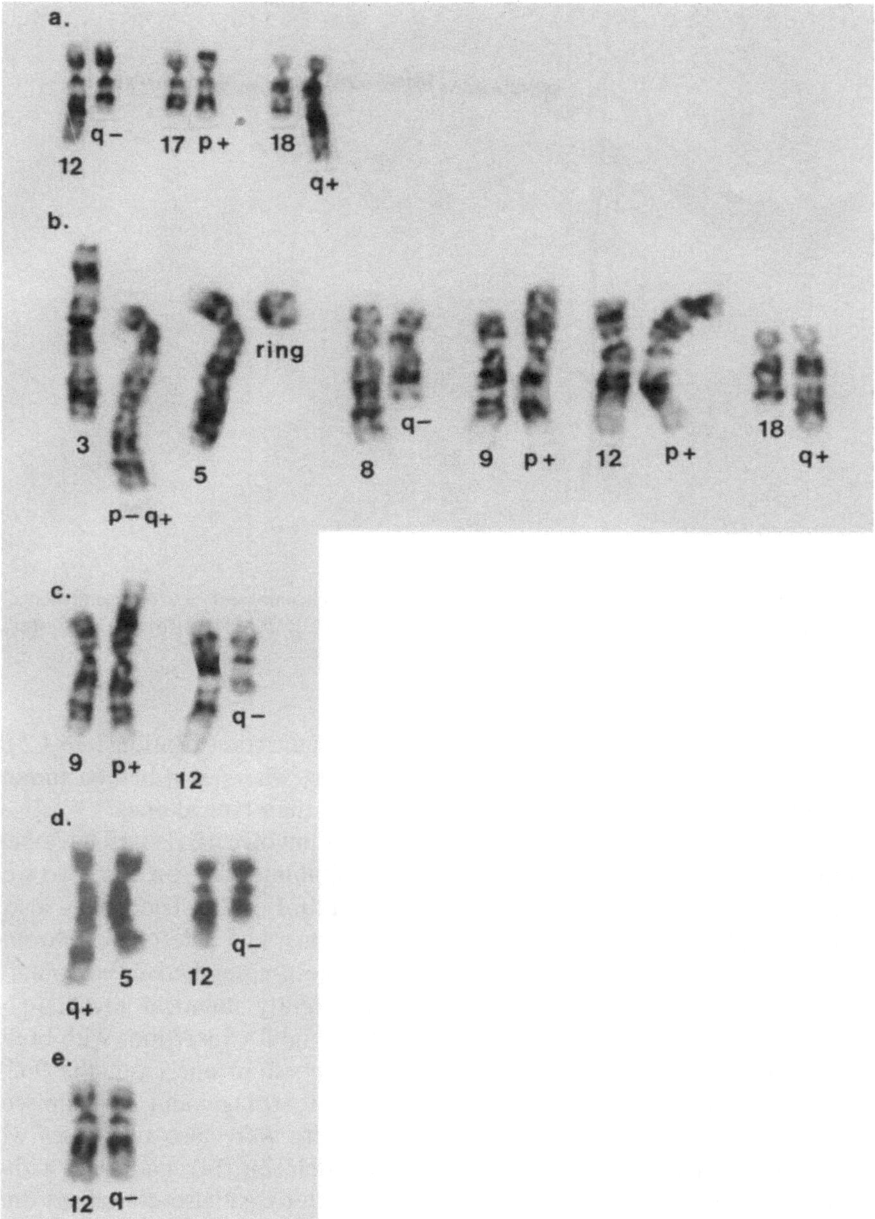

Fig. 2. Partial stemline karyotypes showing various chromosome 12 rearrangements in adenomas from the Swedish material. *a* t(12;18;17)(q15;q21.3;p12); *b* der(3)t(3;5;8;9)(9pter→9p24::3p14→3q25::5q13→5q21::8q12→8qter), r(5)(p15q13), der(8)t(3;8)(q25;q12), der(9)t(9;18)(p24;q12), der(12)t(5;12)(q21;p13), der(18)t(3;18)(p14;q12); *c* t(9;12)(p13−22;q13−15); *d* t(5;12)(q33;q14); *e* del(12)(q13q15)

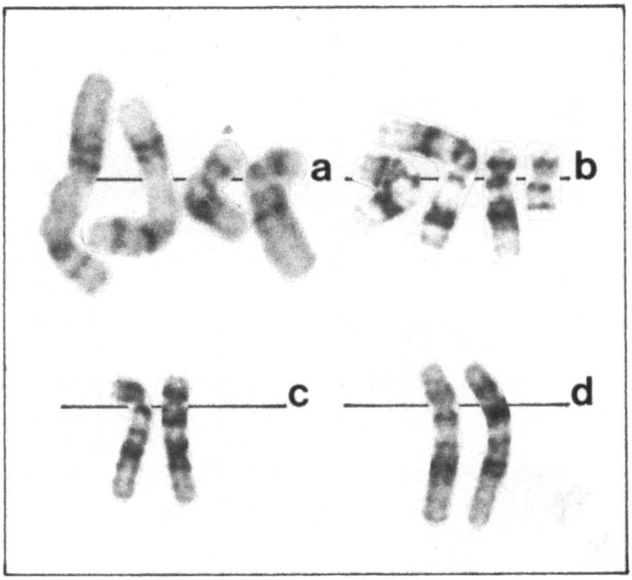

Fig. 3. Partial stemline karyotypes showing various chromosome 12 rearrangements in adenomas from the German material. *a* t(1;12)(p13;q13); *b* t(11;12)(p15;q15); *c* ins(12)(q22q13q15); *d* inv(12)(q14–15q23–24)

abnormalities consists in the frequency of variant translocations. In CML, they only occur in about 5%–10% of all cases, whereas in benign tumors, variant translocations are much more frequent than typical ones.

Chromosome 12 translocations that do not involve 12q13–15 have been found in five cases. In four of these, the breakpoints were on 12p: two were located at p13 (Fig. 2), the other two, at 12q21.3–22. There was also a group of adenomas showing inversions, insertions, and deletions of chromosome 12 instead of translocations. Most of these cases had breakpoints at 12q13–15. Four cases have shown an apparently identical inv(12)(q14–15q23–24) and two cases an inv(12)(p13q13) (Fig. 3). Insertions with breakpoints at 12q13–15 were observed in five tumors. In one case, the 9p21–22→p24 region was inserted into 12q15. This rearrangement is reminiscent of the recurrent t(9;12)(p12–22;q13–15). There were also two cases with deletions of chromosome 12, one with a del(12)(q14) and one with a del(12)(q13q15) (Fig. 2). Interestingly, the latter case also contained dmin chromosomes (Fig. 5) that were suggested to have originated from the deleted 12q segment (Mark et al. 1982).

The pattern of 12q13–15 rearrangements differs in several respects from that observed in pleomorphic adenomas with 8q12 abnormalities. Firstly, tumors with 8q12 changes occur much more frequently than those with 12q13–15 changes. Secondly, the former show a higher frequency of the recurrent t(3;8)(p21;q12) and a corresponding lower frequency of variant

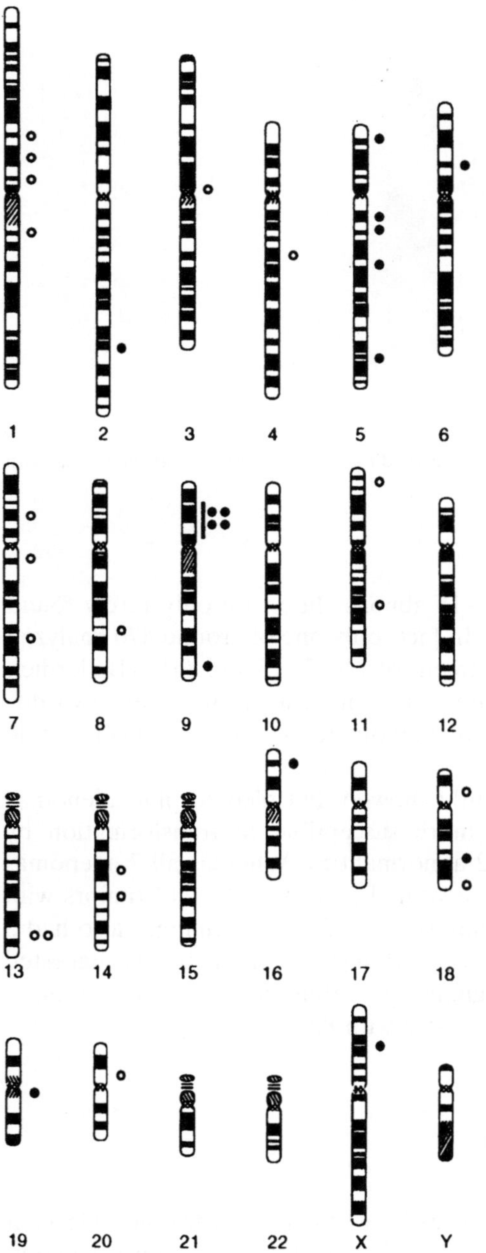

Fig. 4. Ideogram showing the breakpoints on the various translocation partners recombined with 12q13–15→qter in 34 pleomorphic adenomas. The *vertical line* to the *right* of chromosome 9 indicates that the breakpoints have been mapped to the segment 9p12–22 rather than to an individual band. Swedish cases are indicated by *filled circles* and German cases by *open circles*

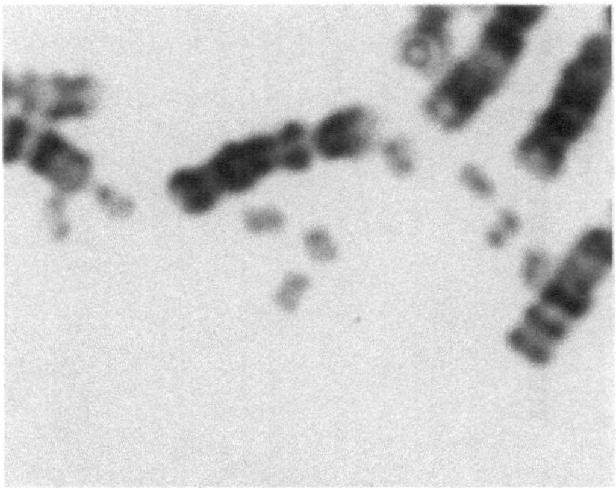

Fig. 5. Partial metaphase showing the presence of dmin chromosomes in an adenoma with del(12)(q13q15)

translocations. Interestingly, 12q13–15 abnormalities are only rarely found together with 8q12 rearrangements. In fact, only one of around 470 analyzed tumors showed simultaneous involvement of 12q13–15 and 8q12 (Bullerdiek et al., unpublished data). These anomalies thus usually represent two different, mutually exclusive patterns of karyotypic evolution in pleomorphic adenomas.

Kazmierczak et al. (1992) recently showed that pleomorphic adenomas with 12q13–15 abnormalities are more susceptible to transformation by SV40 than are adenomas with 8q12 abnormalities. Whereas all 7 adenomas with 12q13–15 changes showed transformed foci, only 4 of 13 tumors with 8q12 changes were transformed. Tumors with 12q13–15 changes also had a higher immortalization rate than those with 8q12 changes. It was suggested that this might reflect a higher malignancy potential on the part of adenomas with 12q13–15 changes than that of those with 8q12 changes.

4 Geographical Heterogeneity

There are several interesting differences between the Swedish and German materials with regard to chromosome 12 abnormalities. The most common recurrent translocations, i.e., t(9;12) and t(5;12), have been found exclusively in the Swedish material, while the less common recurrent t(1;12) and t(12;13) have only been observed in the German material. The frequency of inversions and insertions was much higher in the German material than in the Swedish. The reason for this geographical heterogeneity is uncertain, but it could be attributed to genetic and/or environmental factors. For several reasons, we believe that in pleomorphic adenomas environmental factors are more important determinants for the chromosomal pattern than are genetic factors. This interpretation is also in keeping with data obtained from experimentally-induced tumors, where histopathologically indistinguishable tumors induced by different carcinogenic agents have shown different karyotypic profiles (references in Levan 1975). Cytogenetic heterogeneity between different geographical areas has also been observed, particularly in hematologic neoplasms (Johansson et al. 1991).

5 Molecular Analysis of the 12q13–15 Breakpoint

Recently, several new genes have been isolated and mapped to 12q13–15 (Craig et al. 1992). These include both oncogenes and other growth-related genes of potential interest in relation to tumors. We have previously analyzed the two oncogenes *GLI* and *WNT1*, both located at 12q13, and shown that neither of them are altered in pleomorphic adenomas (Rommel 1991; Sahlin et al. 1992). Nor did we observe any loss of *GLI* alleles in tumor DNA from 13 informative adenomas. These findings strongly indicate that *GLI* and *WNT1* are not involved in 12q13–15 rearrangements in pleomorphic adenomas.

We have now extended these studies to analyses of six additional candidate genes located within the 12q13–15 breakpoint clustering region, i.e., *HOX3* (12q12–13), *COL2A1* (12q13), *ERBB3* (12q13), *A2MR* (12q13.3), *SAS* (12q13.3), and *RAP1B* (12q14) (Craig and McBride 1991; Craig et al. 1992). Using Southern blot analysis, we found no evidence of rearrangement, deletion, or amplification of any of these genes in adenomas with or without 12q13–15 alterations (Fig. 6) (Sahlin et al. and Bullerdiek et al., unpublished observations). These data indicate that neither of these genes seem to be primary targets of rearrangement in adenomas with 12q13–15 abnormalities. Our analyses do not, however, exclude the possibility of rearrangements having occurred outside the DNA sequences analyzed. The present findings concur with recent molecular studies of several other tumor types with rearrangements of 12q13–15, including lipomas, myxoid liposarcomas, and leiomyomas, where one or more of these genes were analyzed and no aberrant restriction pattern was found (reviewed in Stenman et al. 1994; Åman et al. 1992).

Recent studies using fluorescence in situ hybridization (FISH) of established cell lines from myxoid liposarcoma, pleomorphic adenoma, lipoma and leiomyoma with various 12q13–15 abnormalities have shown that the breakpoints in these cell lines are located within a 7 cM region flanked by the markers D12S19 and D12S8, and that the breakpoint in myxoid liposarcoma is proximal to the breakpoints in the three benign tumors (Schoenmakers et al. 1994). An important question in this context is whether

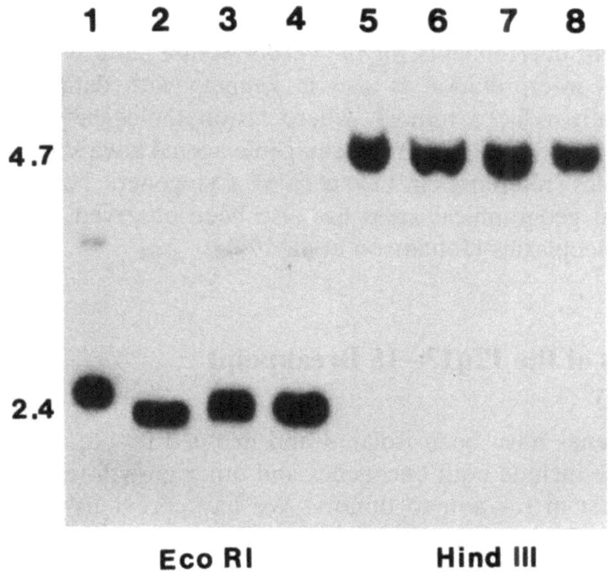

Fig. 6. Restriction fragment analysis of tumor DNAs from four pleomorphic adenomas with translocation breakpoints at 12q13–15. DNAs were digested with EcoRI (*lanes 1–4*) and HindIII (*lanes 5–8*) and hybridized with an *ERBB3* probe. The approximate sizes in kb of the EcoRI and HindIII germline fragments are indicated to the *left*

the three benign tumors might have a common breakpoint. We must await molecular cloning of the various translocation breakpoints for final resolution of this question.

As previously mentioned, we have observed a few adenomas with deletions of 12q13–15. One of these cases contained dmin chromosomes whose origin we suggested as being a deleted segment within 12q13–q15 (Mark et al. 1982). The *SAS* gene, which maps within this segment, was considered a possible candidate gene for the amplification, since *SAS* was recently shown to be amplified in certain soft tissue tumors (Meltzer et al. 1991). However, we were unable to detect amplified *SAS* sequences in this tumor either by conventional Southern analysis or by differential polymerase chain reaction (PCR) analysis (Meltzer et al., unpublished data).

A human papilloma virus (HPV) integration site has been localized to 12q13 (Popescu et al., 1987). Since viral integration is probably nonrandom and may lead to the destabilization and breakage of chromosomes as well as to the activation of oncogenes (Cannizzaro et al., 1988), we analyzed pleomorphic adenomas with 12q13–15 abnormalities for rearrangements of this integration site. Southern analysis revealed a normal restriction pattern in all four tumors studied (Rommel 1991).

To further characterize adenomas with the recurrent t(9;12)(p12–22;q13–15), we examined whether the *INFA* and/or *INFB1* genes, located at 9p22, were altered by the translocation. Based on molecular and/or cytogenetic evidence of deletions in leukemias (Diaz et al. 1988) and high-grade glial tumors (James et al. 1991), these genes have been implicated as tumor suppressor genes. Using restriction fragment analysis, however, we were unable to detect deletions or alterations in any of the two interferon genes, which suggests that they are not involved in the t(9;12) in pleomorphic adenomas (Sahlin et al., unpublished data).

6 Concluding Remarks

Extensive cytogenetic studies of approximately 470 pleomorphic adenomas have unravelled a very specific pattern, especially of translocations with consistent breakpoints at 3p21, 8q12, and 12q13–15. In this paper we have focused on the specificity and implications of chromosome 12 abnormalities. Although the molecular nature of these rearrangements still remains unknown, the study of additional genes in the 12q13–15 region continues to be highly motivated, as this area is a hot spot for chromosomal recombination in three of the cytogenetically most thoroughly studied benign tumor types, i.e., pleomorphic adenoma, lipoma, and leiomyoma. It can be hoped that these studies will disclose essential and critical processes in the development of benign tumors. These mechanisms might prove to be related or perhaps even identical in several types of benign tumors.

Acknowledgements. We would like to thank Paul Meltzer, Matthias Kraus, Ola Myklebost, Erik Lundgren, and Eero Vuorio for DNA probes. This study was supported by grants from the Inga Britt and Arne Lundberg Research Foundation, the Göteborg Medical Society, the Assar Gabrielsson Research Foundation, the Swedish Society for Medical Research, the Swedish Cancer Society, the Deutsche Forschungsgemeinschaft, and the Tönjes-Vagt-Stifung.

References

Åman P, Ron D, Mandahl N, Fioretos T, Heim S, Arheden K, Willén H, Rydholm A, Mitelman F (1992) Rearrangement of the transcription factor gene CHOP in myxoid liposarcoma with t(12;16)(q13;p11). Genes Chrom Cancer 5:278–285

Bullerdiek J, Bartnitzke S, Weinberg M, Chilla R, Haubrich J, Schloot W (1987) Rearrangements of chromosome region 12q13–15 in pleomorphic adenomas of the human salivary gland (PSA). Cytogenet Cell Genet 45:187–190

Bullerdiek J, Wobst G, Meyer-Bolte K, Chilla R, Haubrich J, Thode BU, Bartnitzke S (1993) Cytogenetic subtyping of 220 salivary gland pleomorphic adenomas – correlation to occurrence, histologic subtype, and in vitro celluar behaviour. Cancer Genet Cytogenet 65: 27–31

Cannizzaro LA, Dürst M, Mendez MJ, Hecht BK, Hecht F (1988) Regional chromosome localization of human papilloma virus integration sites near fragile sites, oncogenes, and cancer chromosome breakpoints. Cancer Genet Cytogenet 33:93–98

Craig IW, McBride OW (1991) Report of the committee on the genetic constitution of chromosome 12. Human Gene Mapping 11 (1991). Eleventh international workshop on human gene mapping. Cytogenet Cell Genet 58:555–579

Craig I, Gemmill R, Kucherlapati R (1992) Report of the first international workshop on human chromosome 12 mapping. Cytogenet Cell Genet 61:243–262

Crozat A, Aman P, Mandahl N, Ron D (1993) Fusion of CHOP to a novel RNA-binding protein in human myxoid liposarcoma. Nature 363:640–644

Dalla-Favera R, Neri A, Cesarman E, Lombardi L (1987) Advances on the molecular pathogenesis of Burkitt lymphoma. In: Gale RP, Golde DW (eds) UCLA Symposium on recent advances in leukemia and lymphoma. Liss, New York, pp 165–177

Diaz MO, Aiemin S, LeBeau MM, Pitha P, Smith SD, Chilcote RR, Rowley JD (1988) Homozygous deletion of the alpha- and beta 1-interferon genes in human leukemia and derived cell lines. Proc Natl Acad Sci USA 85:5259–5263

Dubé I, Dixon J, Beckett T, Grossman A, Weinstein M, Benn P, McKeithan T, Norman C, Pinkerton P (1989) Location of breakpoints within the major breakpoint cluster region (bcr) in 33 patients with bcr rearrangement-positive chronic myeloid leukemia (CML) with complex or absent Philadelphia chromosomes. Genes Chrom Cancer 1:106–111

Groffen J, Hermans A, Grosveld G, Heisterkamp N (1989) Molecular analysis of chromosome breakpoints. In: Cohen WE, Moldave K (eds) Progress in nucleic acid research and molecular biology. Academic, New York, vol 36, pp 281–300

James DC, He J, Carlbom E, Nordenskjöld M, Cavenee WK, Collins VP (1991) Chromosome 9 deletion mapping reveals interferon A and interferon B-1 gene deletions in human glial tumors. Cancer Res 51:1684–1688

Johansson B, Mertens F, Mitelman F (1991) Geographic heterogeneity of neoplasia associated chromosome aberrations. Genes Chrom Cancer 3:1–7

Kazmierczak B, Thode B, Bartnitzke S, Bullerdiek J, Schloot W (1992) Pleomorphic adenoma cells vary in their susceptibility to SV40 transformation depending on the initial karyotype. Genes Chrom Cancer 5:35–39

Levan G (1975) Cytogenetic studies in experimental rat sarcomas. Thesis, University of Lund

Mark J, Levan G, Mitelman F (1972) Identification of the G chromosome lost in human meningiomas. Hereditas 71:163–168

Mark J, Dahlenfors R, Ekedahl C, Stenman G (1980) The mixed salivary gland tumor – a normally benign human neoplasm frequently showing specific chromosomal abnormalities. Cancer Genet Cytogenet 2:231–241

Mark J, Dahlenfors R, Ekedahl C (1982) On double-minutes and their origin in a benign human neoplasm, a mixed salivary gland tumor. Anticancer Res 2:261–264

Mark J, Havel G, Grepp C, Dahlenfors R, Wedell B (1990) Chromosomal patterns in human benign uterine leiomyomas. Cancer Genet Cytogenet 44:1–13

Meltzer PS, Jankowski SA, Dal Cin P, Sandberg AA, Paz IB, Coccia MA (1991) Identification and cloning of a novel amplified DNA sequence in a human malignant fibrous histiocytoma derived from a region of chromosome 12 frequently rearranged in soft tissue tumors. Cell Growth Differentiation 2:1–7

Nilbert M, Heim S (1990) Uterine leiomyoma cytogenetics. Genes Chrom Cancer 2:3–13

Pedeutour F, Riviere M, Bouchot C, Mugneret F, Myklebost O, Turc-Carel C (1992) Chromosomal mapping of genes and anonymous DNA segments on human chromosome 12 with respect to the 12q13–q14 nonrandom breakpoint region involved in adipose tissue tumors. Cytogenet Cell Genet 61:260–261

Popescu NC, Amsbaugh SC, Di Paolo JA (1987) Human papilloma virus type 18 DNA is integrated at a single chromosome site in cervical carcinoma cell line SW756. J Virol 51:1682–1685

Rommel B (1991) Molekulargenetische Untersuchungen zum Einfluss bzw. zur Aktivierung von Proto-Oncogenen als Folge instabiler Chromatinkonfigurationen bei benignen Tumoren. Thesis, University of Bremen

Sahlin P, Mark J, Stenman G (1992) INT1 and GLI genes are not rearranged or amplified in benign pleomorphic adenomas with chromosome abnormalities of 12q13-15. Cancer Genet Cytogenet 58:85-88

Sandberg AA (1990) The chromosomes in human cancer and leukemia, 2nd edn. Elsevier, New York

Sandros J, Stenman G, Mark J (1990) Cytogenetic and molecular observations in human and experimental salivary gland tumors. Cancer Genet Cytogenet 44:153-167

Schoenmakers E, Kools P, Mols R, Kazmierczak B, Bartnitzke S, Bullerdiek J, Dal Cin P, De Jong PJ, Van Den Berghe H, Van de Ven W (1994) Physical mapping of chromosome 12q breakpoints in Lipoma, pleomorphic salivary gland adenoma, uterine leiomyoma and myroid Liposarcoma. Genomics 20:210-222

Seifert G, Miehlke A, Haubrich J, Chilla R (1986) Diseases of the salivary glands. Thieme, Stuttgart

Sreekantaiah C, Leong SPL, Karakousis CP, McGee DL, Rappaport WD, Villar HV, Neal D, Fleming S, Wankel A, Herrington PN, Carmona R, Sandberg AA (1991) Cytogenetic profile of 109 lipomas. Cancer Res 51:422-433

Stenman G, Mark J, Ekedahl C (1984) Relationships between chromosomal patterns and proto-oncogenes in human benign mixed salivary gland tumors. Tumour Biol 5:103-117

Stenman G, Sahlin P, Angervall L, Meltzer P, Kindblom L-G (1994) Cytogenetic and molecular studies of clear-cell sarcoma of tendons and aponeuroses. In: Bullerdiek J, Bartnitzke S, Dal Cin P, van de Ven W (eds) Cytogenetics and molecular genetics of chromosome 12 in human solid tumors. Springer, Berlin Heidelberg New York, in press

Turc-Carel C, Limon J, Dal Cin P, Rao U, Karakousis C, Sandberg AA (1986) Cytogenetic studies of adipose tissue tumors. II. Recurrent reciprocal translocation t(12;16)(q13;p11) in myxoid liposarcoma. Cancer Genet Cytogenet 23:291-299

Zankl H, Zang KD (1972) Cytological and cytogenetical studies on brain tumors. IV. Identification of the missing G-chromosome as no. 22 by the fluorescence technique. Hum Genet 14:167-169

Zucman J, Delattre O, Desmaze C, Plougastel B, Joubert I, Melot T, Peter M, De Jong P, Rouleau G, Aurias A, Thomas G (1992) Cloning and characterization of the Ewing's sarcoma and peripheral neuroepihelioma t(11;22) translocation breakpoints. Genes Chrom Cancer 5:271-277

Involvement of Chromosome 12 in Uterine Leiomyoma

R. Vanni[1] and P. Dal Cin[2]

1 Introduction

Leiomyomas are the most common benign mesenchymal tumors of the uterus. They occur mostly in women older than 30 years, less frequently after menopause, and very rarely below the age of 18. Their pathogenesis is not clearly understood, but a role for sex steroid hormones is likely since they increase in size during pregnancy or after administration of oral contraceptives (Buttramn and Reiter 1981) and regress after menopause and under luteinizing hormone-releasing hormone agonist treatment (Benagiano 1989).

Uterine leiomyomas are generally asymptomatic, and only those that create clinical problems are excised. The clinical presentation depends on the size, (ranging from small proliferations to large masses), location (submucosal, intramural, and subserosal), and the number of tumors (as many as 225 have been counted in a single uterus; Lapam and Salomon 1979). Despite of the variety of locations they usually appear as spherical, well-circumscribed gray-white masses, with whorled trabecular pattern with or without pedunculum. Degenerative changes, hemorrhage, and necrosis, can occur, especially in large leiomyomas.

Microscopically they consist of whorled fascicles of spindle-shaped smooth muscle cells with elongated nuclei, finely dispersed chromatin, and small nucleoli (Fig. 1). Mitotic figures are rare (less than 5/10 HPF), and there is little or no cytological atypia. Rare histologic subgroups of leiomyomas can be distinguished, such as cellular, atypical, epithelioid, myxoid, vascular, tubular, and lipogenic types (Zaloudek and Norris 1987).

Determining suitable criteria of malignancy in smooth muscle tumors is still a controversial matter, except in the uterus where the presence of mitotic figures is the main criterion for differential diagnosis between leiomyoma and leiomyosarcoma. (Uterine lesions having between 5 and 9 mitoses/10 HPF are considered borderline lesions). However, uterine leiomyomas are strictly benign and do not undergo malignant transformation.

---

[1] Istituto di Biologia Generale, Universitá degli Studi, Cagliari, Italy, and [2] Centre for Human Genetics, University of Leuven, Leuven, Belgium

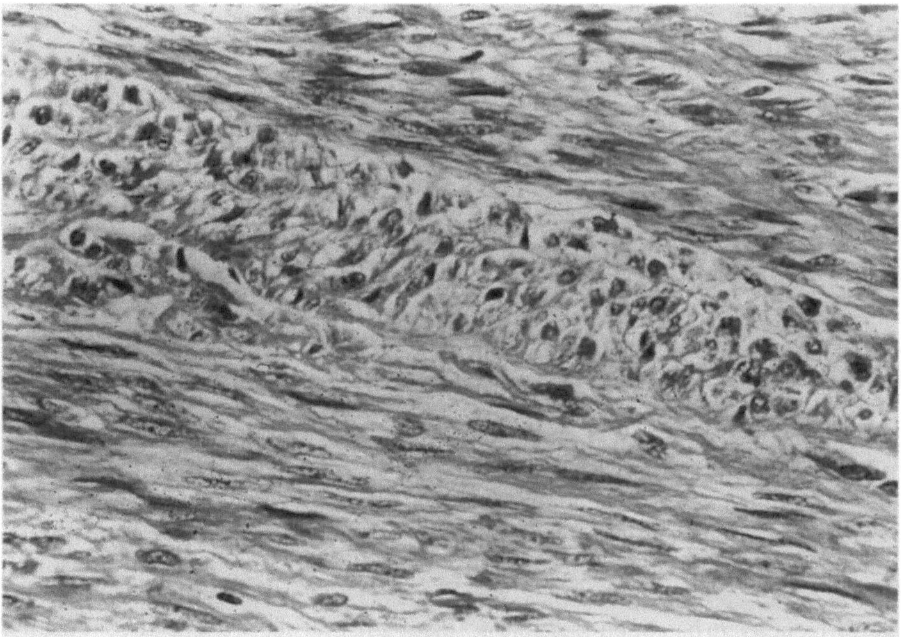

Fig. 1. Whorled anastomosing fasicles of uniform, fusiform, smooth muscle cells

This chapter reviews the published data on the involvement of chromosome 12 in uterine leiomyoma.

2 The Cytogenetic Profile of Uterine Leiomyoma

In 1988 the pioneer studies on uterine leiomyoma demonstrated the presence of single, clonal chromosome changes, mainly involving the 12q14–15 band (Gibas et al. 1988; Heim et al. 1988; Turc-Carel et al. 1988; Vanni and Lecca 1988). Since then, 652 out of 750 (86.9%) tumors have been successfully investigated (Mugneret et al. 1988; Boghosian et al. 1988; Nilbert et al. 1988a,b, 1989a,b, 1990a,b; Sait et al. 1989; Vanni et al. 1989, 1990, 1991; Havel et al. 1989a,b; Kiechle-Schwarz et al. 1990, 1991a,b; Fan et al. 1990; Pandis et al. 1990a,b,c, 1991; Fletcher et al. 1990; Meloni et al. 1991; Stern et al. 1991; Mark et al. 1988, 1990, 1991; Hu and Surti 1991; Bui and Iselius 1991). A normal female karyotype, 46,XX, was found in 69.1% of the tumors. The use of a specific monoclonal antibody (anti-desmin) confirmed that the cells growing in culture were of smooth muscle type (Fig. 2). Among the successfully analyzed tumors 201 cases (30.8%) revealed the presence of clonal chromosome changes. The percentage of abnormal cases reported in the literature varies from series to series, possibly depending

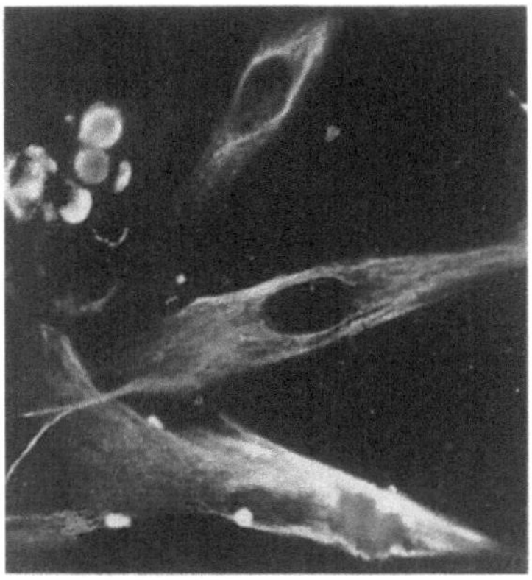

Fig. 2. Positive reaction of cells from leiomyoma culture to the fluorescent antidesmin monoclonal antibody

on different methods used for collecting the samples, preparing the cell-suspension from the specimen, and/or tissue-culture media. The findings indicate a certain degree of cytogenetic heterogeneity in this type of tumor, encompassing cases with normal karyotype, cases with single chromosome change, and cases with complex karyotypes. However, within this apparent heterogeneity it has been possible to distinguish at least five cytogenetically abnormal subgroups, characterized by one of the following clonal abnormalities: (a) involvement of regions 12q14–15 and 14q23–24 (22.4%), mainly as t(12;14)(q14–15; q23–24) (13.5%) (Table 2), (b) del(7)(q21.2q31.2) (19%), (c) trisomy 12 (8%), (d) rearrangements of 6p21-pter (7%), (e) t(1;2)(p36;p24) (3.5%; Fig. 3, Table 1. Note that cases characterized by "other changes (one of the other subgroups)" and cases in parenthesis are not included in the percentage calculation).

3 Involvement of Chromosome 12

3.1 Structural Rearrangements of 12q

The classification of uterine leiomyoma in five cytogenetically abnormal subgroups does not imply that within each group changes are identical. Each subgroup may show "variant changes," in which, however, characteristic breakpoints are always involved. Considering the cases in which chromosome 12 is structurally involved, and summing up the tumors characterized by the 12;14-translocation and the tumors in which the region 12q14–15 is involved in changes other than t(12;14), it turns out that the 27% of the abnormal cases is associated with chromosome 12q14–15 rearrangements. Within this group the t(12;14), as the only change, represents the 23.6% of the cases and is regarded as a primary change, i.e., a change considered to be of importance in tumor initiation (Fig. 4). In the earliest studies it was found both as single change or together with other few chromosome abnormalities leading to a

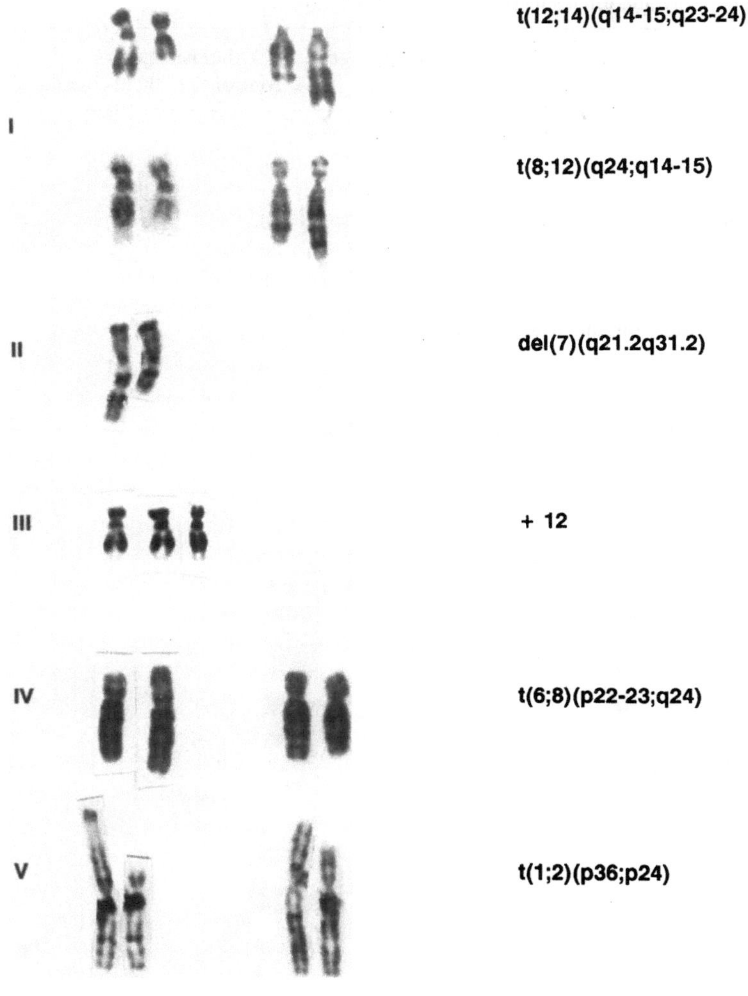

Fig. 3. Partial karyotypes representative of the five largest cytogenetically abnormal groups of uterine leiomyoma

debate as to whether the critical breakpoint was on chromosome 12 or 14. The finding of the first "variant translocation," a t(X;12)(p22.3q15) (Vanni and Lecca 1988), indicated the 12q14–15 breakpoint as the more critical one. On the other hand, the finding of a case with a dir ins(14;6)(q23;p23p25) (Mugneret et al. 1988) suggested a primary role for the 14q23–24 breakpoint. In the light of the 652 so far examined cases, the first interpretation seems to be likely, since a number of tumors have been reported in which 12q14–15 band is affected without obvious involvement of the 14q. However, it seems as if the 12q14–15 breakpoint can be involved in different rearrangements, as reported in Table 2 (Fig. 5). Reciprocal (12;14)-translocations or insertions (Heim et al. 1988; Turc-Carel et al. 1988), inv(12) involved in the trans-

Table 1. Cytogenetically abnormal subgroups in uterine leiomyoma

Subgroup	Only change[a]	+ Other change(s)	+ Other change (one of the other subgroups)[b]
12q14–15 involvement	19	26	10
7q-	39 (5)	4	8
+12	15 (2)	3	1
6p21→ter involvement	9 (2)	7	4
t(1;2)	5	2	–
Varia	44 (1)	22 (2)	–

[a] In parenthesis: in cases with only change in a second clone.
[b] Total number of cases is 12.

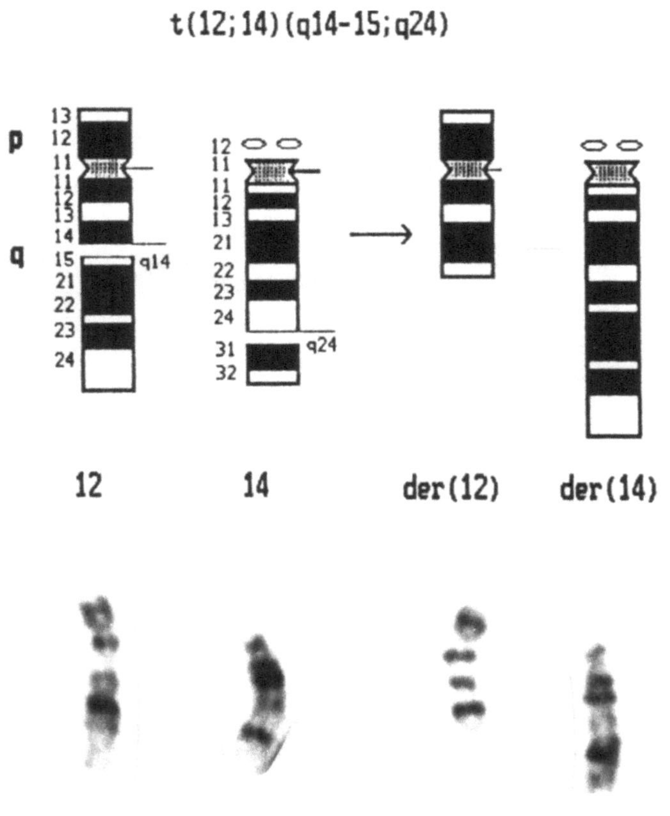

Fig. 4. Diagram and corresponding partial karyotype showing the classical involvement of 12q14–q15 region:t(12;14)(q14–15;q23–24)

Table 2. Rearrangements of the 12q14–15 region in uterine leiomyoma

t(12;14) only change[a]	12;14 + other change(s)	t(12;?) only change[b]	t(12;?) + other change(s)	t(12;14;?)[c]	der(12) and/ or der(14)
13	20 (5)[d]	6	7 (2)[d]	4 (2)[d]	5 (1)[d]

[a] Including t[inv(12;14);14] or ins(12;14).
[b] Including inv(12), 12q-.
[c] Including der(12), der(14).
[d] Plus abnormality of one of the other subgroups.

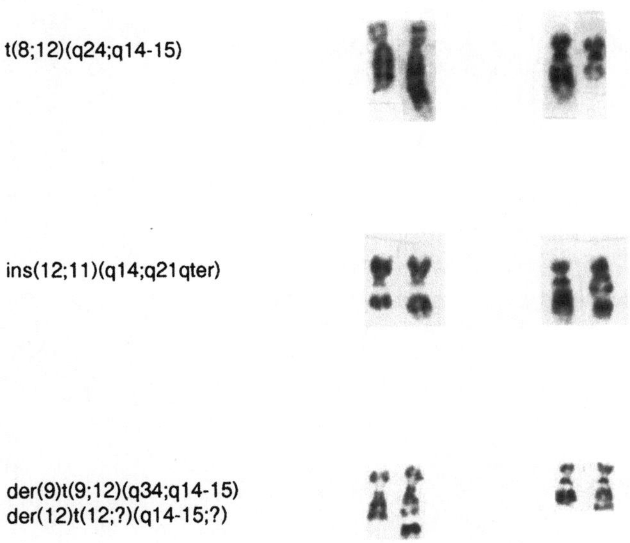

Fig. 5. Partial karyotypes showing other type of involvement of the 12q14–15 region in uterine leiomyoma

location (12;14) (Mark et al. 1988), "variant translocations," i.e., without obvious involvement of band 14q23–24 (Vanni and Lecca 1988), and three-way (complex) translocations (Hu and Surti 1991) are examples of different involvements of 12q14–15. This situation strikingly recalls the cytogenetic pattern of chronic myeloid leukemia, where, besides the standard t(9;22)(q34;q11) translocation, variant translocations (without apparent involvement of 9q34) and complex translocations can be found. A reciprocal involvement of 12q and 14q at molecular level, even in the cases without visible changes in 14q, should be taken into account when considering the problem of which gene(s), is critical in this subgroup. Comparing the most frequent cytogenetic subgroups of uterine myomas, the one characterized by the 12q14–15 change and the one with 7q-deletions, it seems that cases with 7q-, as the only

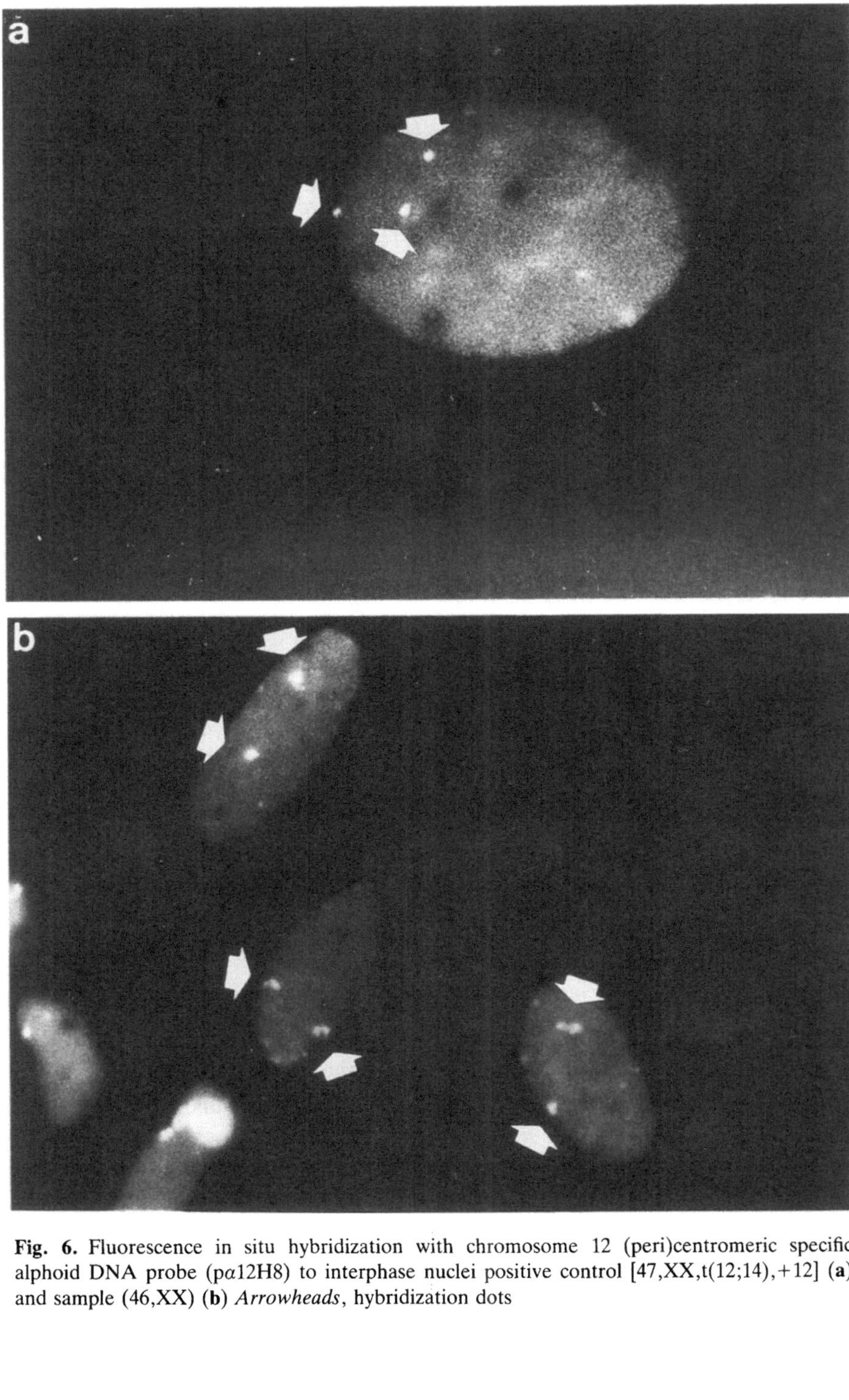

Fig. 6. Fluorescence in situ hybridization with chromosome 12 (peri)centromeric specific alphoid DNA probe (pα12H8) to interphase nuclei positive control [47,XX,t(12;14),+12] (**a**) and sample (46,XX) (**b**) *Arrowheads*, hybridization dots

change, are more frequent than the cases with the sole rearrangement of the 12q14–15 band (Table 1). However, when considering the subgroups irrespective of the presence of other changes, the cases where the 12q14–15 band is involved are slightly more frequent (22.3%) than the cases with 7q deletions (19%) (Table 1).

3.2 Numerical Aberrations of Chromosome 12

Trisomy 12 characterizes another subgroup of uterine myomas. It has been reported in almost 10% of the abnormal cases. It was the only change in 13 of the 19 cases in which it was present. A feature of this cytogenetic subgroup is the absence of complex karyotypes, moreover no more than one additional change has so far been found together with trisomy 12. The presence of this aberration has been also tested by interphase cytogenetics, using a chromosome 12 alphoid DNA probe (pα12H8). The study demonstrated that trisomy 12 is not underscored in cases in which a normal karyotype was determined by conventional cytogenetics (Fig. 6; Vanni et al. 1992).

In conclusion, uterine leiomyomas are characterized by numerical or structural abnormalities of chromosome 12, (12q14–15 involvement, and trisomy 12) in 30% of cytogenetically abnormal cases. This finding seems to indicate a key role of 12q14–15 region in the pathogenesis of one and perhaps two subgroups of uterine leiomyoma. The group of tumors apparently without chromosome changes will constitute a challenge for molecular studies once the underlying changes have been identified. The preferential involvement of the long arm of chromosome 12, however, is not an exclusive feature of this tumor. Other benign proliferations, such as lipomas and pleomorphic adenoma of salivary glands (Sreekantaiah and Sandberg 1991), exhibit consistent rearrangements in 12q14–15 in one of their cytogenetically abnormal subgroups, providing important evidence that this specific region of chromosome 12 is a candidate for the presence of key genes implicated in the proliferation process. In fact, rearrangements of 12q13–q15 have been also described in malignant tumors such as myxoid liposarcoma, clear cell sarcoma, uterine leiomyosarcoma, and chondrosarcoma (Mitelman 1991). Different, perhaps closely related genes or a unique gene, may be implicated in both the benign and the malignant.

Acknowledgements. The authors wish to thank Prof. H. Van den Berghe for helpful discussion and Rita Logist for clerical assistance.

References

Benagiano G (1989) Uterine fibroids: literature review and summary of posters. Horm Res 32 [Suppl 1]:120–124
Boghosian L, Dal Cin P, Sandberg AA (1988) An interstitial deletion of chromosome 7 may characterize a subgroup of uterine leiomyoma. Cancer Genet Cytogenet 34:207–208
Bui TH, Iselius L (1992) t(7;22)(q31;12) in a uterine leiomyoma. Cancer Genet Cytogenet 59:219
Buttramn V Jr, Reiter RC (1981) Uterine leiomyoma etiology, symptomatology, and management. Fertil Steril 36:433
Fan S-X, Sreekantaiah C, Berger CS, Medchill M, Pedron S, Sandberg AA (1990) Cytogenetic findings in nine leiomyomas of the uterus. Cancer Genet Cytogenet 47:179–189
Fletcher JA, Morton CC, Pavelka K, Lage JM (1990) Chromosome aberrations in uterine smooth muscle tumors: potential diagnostic relevance of cytogenetic instability. Cancer Res 50:4092–4097
Gibas Z, Griffin CA, Emanuel BS (1988) Clonal chromosome rearrangements in a uterine myoma. Cancer Genet Cytogenet 32:19–24
Havel G, Dahlenfors R, Wedell B, Mark J (1989a) Similar chromosomal evolution in a uterine stromomyosarcoma and in one of two leiomyomas from the same patient. APMIS 97:143–146
Havel G, Wedell B, Dahlenfors R, Mark J (1989b) Cytogenetic relationship between uterine lipoleoimyomas and typical leiomyomas. Virchows Arch [B] 57:77–79
Heim S, Nilbert M, Vanni R, Floderus U-M, Mandahl N, Liedgren S, Lecca U, Mitelman F (1988) A specific translocation, t(12;14)(q14–15;q23–24), characterizes a subgroup of uterine leiomyomas. Cancer Genet Cytogenet 32:13–17
Hu J, Surti U (1991) Subgroups of uterine leiomyomas based on cytogentic analysis. Hum Pathol 22:1009–1016
Kiechle-Schwarz M, Berger CS, Surti U, Sandberg AA (1990) Rearrangement of band 10q22 in leiomyoma and leiomyosarcoma of the uterus. Cancer Genet Cytogenet 47:95–100
Kiechle-Schwarz M, Sreekantaiah C, Berger CS, Pedron S, Medchill MT, Surti U, Sandberg AA (1991a) Nonrandom cytogenetic changes in leiomyomas of the female genitourinary tract – a report of 35 cases. Cancer Genet Cytogenet 53:125–136
Kiechle-Schwartz M, Pfleiderer A, Sreekantaiah C, Berger CS, Medchill MT, Sandberg AA (1991b) Cluster of trisomy 12 to tumors of the female genitourinary tract. Cancer Genet Cytogenet 53:273–275
Lapam B, Salomon L (1979) Diffuse leiomyomatosis of the uterus precluding myomectomy. Obstet Gynecol 53 [Suppl]:82s
Mark J, Havel G, Grepp C, Dahlenfors R, Wedell B (1988) Cytogenetical observations in human benign uterine leiomyomas. Anticancer Res 8:621–626
Mark J, Havel G, Grepp C, Dahlenfors R, Wedell B (1990) Chromosomal patterns in human benign uterine leiomyomas. Cancer Genet Cytogenet 44:1–13
Mark J, Havel G, Grepp C, Dahlenfors R, Wedell B (1991) Cytogenetics of multiple uterine leiomyomas, parametrial leiomyoma and disseminated peritoneal leiomyomatosis. Anticancer Res 11:33–40
Meloni AM, Surti U, Sandberg AA (1991) Deletion of chromosome 13 in leiomyomas of the uterus. Cancer Genet Cytogenet 53:199–203
Mitelman F (1991) Catalog of chromosome aberrations in cancer, 4th edn. Liss, New York
Mugneret F, Lizard-Nacol S, Volk C, Cuiseinier F, Colin F, Turc-Carel C (1988) Association of breakpoint 14p23 with uterine leiomyoma. Cancer Genet Cytogenet 34:201–206
Nilbert M, Heim S, Mandahl N, Flodérus U-M, Willén H, Åkerman M, Mitelman F (1988a) Ring formation and structural rearrangements of chromosome 1 as secondary changes in uterine leiomyomas with t(12;14)(q14–15;q23–24). Cancer Genet Cytogenet 36:183–190

Nilbert M, Heim S, Mandahl N, Flodérus U-M, Willén H, Mitelman F (1988b) Karyotypic rearrangements in 20 uterine leiomyomas. Cytogenet Cell Genet 49:300–304

Nilbert M, Heim S, Mandahl N, flodérus U-M, Baldetorp B, Mitelman F (1989a) Complex karyotypic anomalies in a bizarre leiomyoma of the uterus. Genes Chrom Cancer 1:131–134

Nilbert M, Heim S, Mandahl N, Flodérus U-M, Willén H, Mitelman F (1989b) Different karyotypic abnormalities, t(1;6) and del(7), in two uterine leiomyomas from the same patient. Cancer Genet Cytogenet 42:51–53

Nilbert M, Heim S, Mandahl N, Flodérus U-M, Willén H, Mitelman F (1990a) Characteristic chromosome abnormalities, including rearrangements of 6p, del(7q), +12, and t(12;14), in 44 uterine leiomyomas. Hum Genet 85:605–611

Nilbert M, Heim S, Mandahl N, Flodérus U-M, Willén H, Mitelman F (1990b) Trisomy 12 in uterine leiomyomas. A new cytogenetic subgroup. Cancer Genet Cytogenet 45:63–66

Pandis N, Bardi G, Sfikas K, Panayotopoulos N, Tszrkezoglou A, Fotiou S (1990a) Complex chromosome rearrangements involving 12q14 in two uterine leiomyomas. Cancer Genet Cytogenet 49:51–56

Pandis N, Heim S, Bardi G, Flodérus U-M, Willén H, Mandahl N, Mitelman F (1990b) Parallel karyotypic evolution and tumor progression in uterine leiomyoma. Genes Chrom Cancer 2:311–317

Pandis N, Heim S, Bardi G, Mandahl N, Mitelman F (1990c) High resolution mapping of consistent leiomyoma breakpoints in chromosomes 12 and 14 to 12q15 and 14q24.1. Genes Chrom Cancer 2:227–230

Pandis N, Heim S, Bardi G, Flodérus U-M, Willén H, Mandahl N, Mitelman F (1991) Chromosome analysis of 96 uterine leiomyomas. Cancer Genet Cytogenet 55:11–18

Sait SNJ, Dal Cin P, Ovanessoff S, Sandberg AA (1989) A uterine leiomyoma showing both t(12;14) and del(7) abnormalities. Cancer Genet Cytogenet 37:157–161

Sreekantaiah C, Sandberg AA (1991) Clustering of aberrations to specific chromosome regions in benign neoplasms. Int J Cancer 48:194–198

Stern C, Kazmierczak B, Thode B, Rommel B, Bartnitzke S, Dal Cin P, Van de Ven W, Van den Berghe H, Bullerdiek J (1991) Leiomyoma cells with 12q15 aberrations can be transformed in vitro and show a relatively stable karyotype during precrisis period. Cancer Genet Cytogenet 54:223–228

Turc-Carel C, Dal Cin P, Boghosian L, Terk-Zakarin J, Sandberg AA (1988) Consistent breakpoints in region 14q22–q24 in uterine leiomyoma. Cancer Genet Cytogenet 32:25–31

Vanni R, Lecca U (1988) Involvement of the long arm of chromosome &2 in chromosome rearrangements of uterine leiomyoma. Cancer Genet Cytogenet 32:33–34

Vanni R, Nieddu M, Paoli R, Lecca U (1989) Uterine leiomyoma cytogenetics. I. Rearrangements of chromosome 12. Cancer Genet Cytogenet 37:49–54

Vanni R, Dal Cin P, Van Den Berghe H (1990) Is the chromosome band 1p36 another hot-spot for rearrangements in uterine leiomyoma? Genes Chrom Cancer 2:255–256

Vanni R, Lecca U, Faa G (1991) Uterine leiomyoma cytogenetics. II. Report of fourty cases. Cancer Genet Cytogenet 53:247–256

Vanni R, Van Roy N, Lecca U, Speleman F (1992) Uterine leiomyoma cytogenetics. III. Interphase cytogenetic analysis of karyotypically normal uterine leiomyomas excludes possibility of undetected trisomy 12. Cancer Genet Cytogenet 62:40–42

Zaloudek CJ, Norris HJ (1987) Mesenchymal turmors of the uterus. In: Kurman RJ (ed) Blaustein's pathology of the female tract, 3rd edn. Springer, Berlin Heidelberg New York

Lipoma Cytogenetics

N. Mandahl

1 Introduction

The occurrence of nonrandom clonal cytogenetic abnormalities in tumor cells has been unequivocally demonstrated (Heim and Mitelman 1987; Sandberg 1990). It should, however, be stressed that the data are heavily biased in favor of hematologic neoplasms; of the total data base comprising 14 141 human tumors reported to have clonal chromosome aberrations, 68% involve hematologic disorders, 11%, lymphomas, and 21%, solid tumors (Mitelman 1991). This is highly disproportionate with respect to the relative contribution of these disorders to human cancer morbidity and mortality.

Due to recent methodological improvements in solid tumor cytogenetics, it has become quite clear that at least all extensively investigated tumor types have clonal chromosome abnormalities. This is true not only for malignant tumors, but also for benign tumors, and in fact several of the most extensively investigated solid tumors are benign lesions. These include in particular meningioma, pleomorphic adenoma of the salivary gland, lipoma, and uterine leiomyoma. In general, benign tumors have less complex aberrations than malignant tumors, although exceptions on both sides are known.

For several years now, cytogenetic analysis of hematologic disorders has been widely used in clinical practise. The specificity of the acquired aberrations make them quite useful for diagnosis and monitoring during treatment follow-up. The karyotypic aberrations may also be a useful tool in prognostication. Cytogenetic analysis of solid tumors has as yet played a less important clinical role. There is, however, convincing evidence that the chromosomal rearrangements are pathogenetically important, and karyotypic analysis may reveal which chromosome segments contain genes that are involved in the carcinogenic process. Identification of these gene mutations and then of the corresponding altered gene products may give an insight into how the action and interaction of abnormal macromolecules turn a normally functioning cell into a constitutively proliferating neoplastic cell.

The superiority of cytogenetic analysis over molecular genetic analysis as a screening procedure to detect tumor-associated genetic changes makes

Department of Clinical Genetics, Lund University Hospital, 221 85 Lund, Sweden cytogenetics the natural first choice to identify the primary and secondary chromosome changes that characterize a particular tumor type. Based on these findings, the powerful molecular genetic analysis can be directed to a restricted number of potentially pathogenetically interesting chromosome regions. The following will primarily summarize the cytogenetic findings of benign and borderline adipose tissue tumors and only very briefly the cytogenetic findings in other adipose tissue tumors and other tumors with related chromosomal aberrations.

2 Cytogenetic Methodology

Cytogenetic analysis of human solid tumors has long been hampered by the lack of appropriate methods for obtaining high-quality chromosome preparations. Before the mid 1980s, a substantial proportion of cytogenetic data was obtained from direct preparations and established cell lines. Chromosome preparations made directly from fresh tumor tissue yield frequently condensed chromosomes with suboptimal banding patterns and require tumor tissue with a reasonably high mitotic activity. Hence, many slowly growing benign tumors, for example lipomas, are not suitable for direct preparations. During the latest decade, improved methods for cytogenetic analysis of solid tumors have been developed. These include controlled enzymatic dissociation of tumor tissue, short-term culturing in more and more refined media with growth-promoting additives, and occasionally modified growth substrates (see, for example, Mandahl 1992).

In lipomas, mitotic cells are extremely rarely seen in the histopathological analysis of sections. However, cells from lipoma tissue dissociated in collagenase II for 2–4 h readily grow in basic media without any additives other than fetal calf serum and glutamine (Mandahl et al. 1988; Sreekantaiah et al. 1991c). Mitotic activity of the cells with clonal cytogenetic aberrations usually ceases within a few weeks and cytogenetically normal cells take over. The cultures do not transform spontaneously to permanent cell lines; to achieve this special transfection, particular methods have to be applied (Stern et al. 1991). The cells that attach and grow in the primary cultures when clonal chromosome aberrations are usually found look like small fibroblasts, but it is not known what type of cells they represent. No mature fat cells are seen in the cultures.

3 Cytogenetic Findings

The finding of a normal karyotype in cultured tumor cells is inconclusive, because it is not known whether these represent tumor parenchyma cells with no or cytogenetically undetectable chromosomal aberrations or dividing stromal cells; this question will not be dealt with here. The total data base of published, cytogenetically abnormal benign and borderline malignant adipose tissue tumors consists of 114 tumors. Of these, 89 were typical lipomas, five atypical lipomas, two angiomyolipomas, two lipoblastomas, one fibrolipoma, one myxolipoma, one hibernoma, and 13 well-differentiated liposarcomas. There is some overlap between cytogenetic and histopathologic subgroups, and I have chosen to present the data according to the cytogenetic characteristics.

3.1 Tumors with Rearrangements of Chromosome Segment 12q13–15

A central segment of the long arm of chromosome 12, consisting of bands q13, q14, and q15, is a hot spot for chromosome rearrangements in lipomas (Heim et al. 1986; Turc-Carel et al. 1986a, 1988; Karakousis et al. 1987; Mandahl et al. 1987, 1988; Dal Cin et al. 1988b; Sreekantaiah et al. 1991c; Åman et al. 1992; Debiec-Rychter et al. 1992). The majority of tumors (60) have had at least one breakpoint localized to this segment, which may recombine with a large variety of other chromosomes and chromosome bands; some seem to be preferred, however, in particular the distal portion of the long arm of chromosome 3. In general, the karyotypic changes have been balanced and simple, with only a single aberration giving rise to pseudodiploid karyotypes; occasionally, two or more aberrations have been found in a clone and in some cases two cytogenetically related clones, indicating clonal evolution, were present. The average number of breakpoints per tumor was 3.3.

3.1.1 Translocations Involving 3q and 12q

Reciprocal translocations between chromosomes 3 and 12 with breakpoints within the segments 3q27–28 and 12q13–15, t(3;12)(q27–28;q13–15; Fig. 1a) are the most common recurrent aberrations in lipoma and have been found in 14 tumors, one of which was diagnozed as an atypical lipoma. In ten tumors it was the sole anomaly in at least one clone, indicating that

Fig. 1,A,B. Partial karyotypes from two case of lipoma showing **A** an aberration interpreted as t(3;12) (q27;14), and **B** an ins dup (12;12) (q24;q13q24). *Arrowheads indicate breakpoints*

t(3;12) is of primary pathogenetic importance in lipoma. Rearrangements involving 3q27–28, but not 12q13–15, have not been found, whereas the reverse is quite common (see below). Although the possibility cannot be excluded that tumors without 3q27–28 aberrations have molecular genetic changes of this segment, it seems likely that changes of 12q13–15 rather than those of 3q27–28 are more important in lipomas.

3.1.2 Other 12q13–15 Aberrations

Intrachromosomal as well as a variety of interchromosomal rearrangements involving 12q13–15 have been found in 46 tumors. Only one of these had a variant diagnosis, a renal angiomyolipoma. A single cytogenetically abnormal clone was found in 37 tumors, and five tumors had two related clones, three tumors had three related clones, and one had two unrelated clones. All chromosomes except 9, 14, 16, 18, 20, and Y were involved in rearrangements with 12q13–15, and in seven tumors these involved three or more chromosomes. The average number of breakpoints per tumor was 3.5.

In seven tumors the 12q13–15 aberrations were intrachromosomal (Fig. 1b), including inversions, deletions, duplications, or combinations thereof. Some other bands or segments consisting of two neighboring bands were nonrandomly involved in the aberrations. In order of decreasing frequency these were: 21q21–22 in six tumors, 1p32–33 and 2p22–23 in four tumors each, and 1p36, 5q33, and 11p13 in two tumors each. Thus, in total two-thirds of lipomas with 12q13–15 aberrations have breakpoints within the bands or segments (including 3q27–28) specifically mentioned above.

3.2 Tumors with Aberrations of 6p

Another cytogenetic subgroup characterized by aberrations of the short arm of chromosome 6 has been distinguished in lipomas (Mandahl et al. 1988; Sait et al. 1989; Haus et al. 1990; Sreekantaiah et al. 1991c). There are nine such tumors reported, with the ten breakpoints (because of one duplication) in 6p distributed to bands p21 and p23 (three breaks in each), p22 (two breaks), and p11 and p25 (one break in each). In one tumor there was recombination between 6p and 12q in an ins(6;12)(p23;q15q21). Except for two tumors with t(2;6)(q36;p21) and t(2;6)(q37;p11), there was no similarity between the aberrations other than the 6p involvement. This lack of specificity and the overall distribution of breakpoints in lipomas make it doubtful whether it is relevant and meaningful to distinguish lipomas with 6p aberrations as a particular subgroup. There are two other chromosome segments of similar size as 6p, 1p32–36, and 17q with at least the same level of breakpoint clustering as in 6p (see Fig. 4); admittedly, the former segments are more often affected in tumors with 12q13–15 aberrations than is 6p.

3.3 Tumors with Aberrations of 13q

Another cytogenetic subgroup reported in the literature is characterized by aberrations of the long arm of chromosome 13 (Sandberg et al. 1986; Mandahl et al. 1988; Sreekantaiah et al. 1991c; Åman et al. 1992). Eleven cases are known; three tumors had rearrangements of both 13q and 12q13–15 and one tumor had both 13q and 6p changes, but except for one case with a t(12;13), there was no recombination between 13 and 12 or between 13 and 6. Of the eleven tumors, seven had deletions, three had translocations, and one had an inversion. This is in contrast to the previously mentioned subgroups with 12q13–15 and 6p aberrations, where the majority of the changes are translocations and very few are deletions. The breakpoints in chromosome 13 cluster to two bands, q12 and q22, with nine and six of the 18 breakpoints, respectively. Five tumors had an identical del(13)(q12q22) and the segment 13q13–22 was lost in all seven tumors with deletions. In two tumors, deletion of chromosome 13 was the sole anomaly. Two tumors in this group were not typical lipomas; one was diagnosed as a fibrolipoma and one as an atypical lipoma.

3.4 Tumors with Ring and/or Giant Marker Chromosomes

Supernumerary ring chromosomes and/or usually very large marker chromosomes have been found in 19 tumors (Figs. 2, 3); four were classified as lipomas, three as atypical lipomas, and 12 as well-differentiated liposarcomas (Becher et al. 1984; Heim et al. 1987; Karakousis et al. 1987; Mandahl et al. 1988; Sreekantaiah et al. 1991c, 1992; Åman et al. 1992; Mitchell et al. 1992; Stephenson et al. 1992). Ring chromosomes were more common than giant markers and in some cases they may appear together in the same tumor. The number of rings varied from one to three, and quite often the number and size differed between metaphases from the same tumor. The chromosome number in these tumors ranged from 42 to 51, but the majority of cases were hyperdiploid. Rings and markers were the sole changes in at least one clone in the majority of cases, but in eight tumors different, additional, numerical and/or structural changes were found. The origin of the ring and marker chromosomes has not been possible to identify. Some of the tumors with these aberrations also showed multiple telomeric associations and a relatively high frequency of nonclonal aberrations.

3.5 Tumors with Other Aberrations

Twenty tumors had aberrations that do not fit into any of the previously mentioned cytogenetic subgroups (Sandberg et al. 1986; Mandahl et al. 1987, 1988; Dal Cin et al. 1988a, 1992; De Jong et al. 1988; Peulve et al.

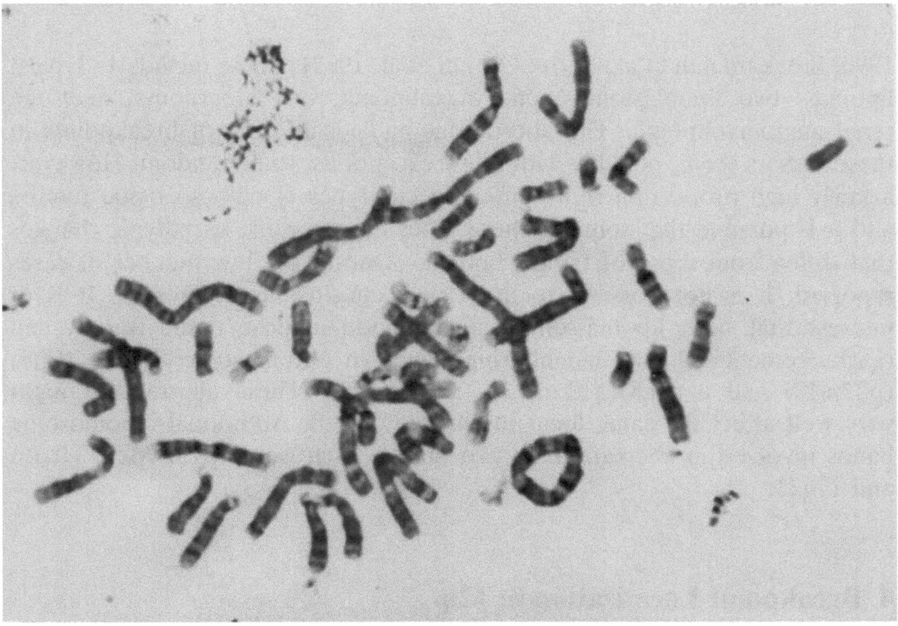

Fig. 2. Metaphase cell from an atypical lipoma with a large, supernumerary ring chromosome

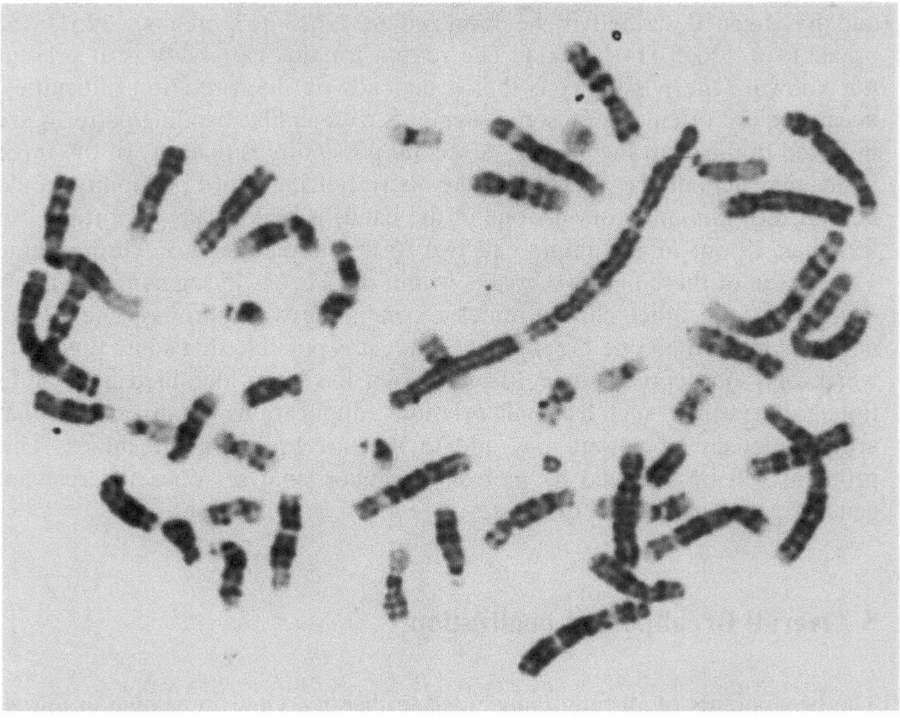

Fig. 3. Metaphase cell from an atypical lipoma showing a giant marker chromosome

1990; Sreekantaiah et al. 1991c; Ohjimi et al. 1992). These include 15 typical lipomas, two lipoblastomas, one myxolipoma, one hibernoma, and one renal angiomyolipoma. The aberrations and localization of breakpoints in these tumors seem, possibly with a few exceptions, to be random. However, a fairly high proportion represented special types of adipose tissue tumors and it is possible that some of these types have specific karyotypic changes that differ from those of typical lipomas. Due to the low number of cases reported, it is not possible to draw such conclusions at present. It is of interest that both lipoblastomas, a childhood adipose tissue tumor, had rearrangements of two neighboring bands in chromosome 8 in a t(7;8)(p22;q12) and a t(3;8)(q12;q11.2), respectively. These aberrations might very well affect the same locus in chromosome 8. Additional chromosome bands involved in aberrations in two tumors each were 1p13, 7p22, 11q13, and 17q21.

4 Breakpoint Localization in 12q

The breakpoint in 12q in the t(3;12) and other aberrations involving the same segment of chromosome 12 appear to vary from one tumor to another and has been described to be localized to 12q13 (23 breaks), 12q14 (32 breaks), or 12q15 (14 breaks). The reason for this variability is at present not known. One possibility is that it depends on interpretation difficulties, because quite often chromosome segments with similar banding patterns are involved in the rearrangements. A second possibility is that any of the three bands may actually be affected by the aberrations. A third possibility would be that recombination of only one of the bands is biologically important. No less than 19 out of 60 tumors had two to four breakpoints in chromosome 12, and ten of these had two breaks within the 12q13–15 segment. With few exceptions, the other chromosomes recombining with chromosome 12 did not have more than one breakpoint. Thus, it is possible that some tumors in which only one chromosome 12 breakpoint has been identified have additional aberrations, such as small deletions, duplications, or inversions, that are not detected but that may add to the interpretation difficulties. This problem will be resolved by molecular genetic analyses once the relevant gene or genes have been isolated.

5 Overall Breakpoint Localization

The breakpoints of all tumors included in this overview are shown in Fig. 4. Apart from the obvious clustering of breakpoints to the three bands in the segment 12q13–15, some other bands have been involved in aberrations five times or more: 1p36, 3q27, 3q28, 11q13, 12p11, 12q24, 13q12, 13q22, and 21q21. Given the interpretation difficulties of the exact breakpoint localization, breakpoints also tend to cluster to some chromosome segments consisting of three neighbouring bands, i.e., 1p32–34, 2p21–23, and 6p21–23. Chromosomes 1, 2, 3, 6, 12, 13, and 17 were involved in rearrangements in more than 10% of the tumors.

6 Cytogenetic Comparisons with Other Tumors

The karyotypic profiles of lipomas share characteristics with those of several other tumor types. Only a short summary of these similarities will be given, because details of some of these tumors will be presented elsewhere in this volume.

6.1 Lipomas and Other Adipose Tissue Tumors

Myxoid liposarcomas are characterized by a t(12;16)(q13;p11) rearrangement (Turc-Carel et al. 1986b; Örndal et al. 1990; Sreekantaiah et al. 1991b). This aberration is highly specific and has been found in more than 90% of the tumors, in more than half of them as the sole anomaly. The transcription factor gene, *CHOP*, which has been shown to be consistently rearranged in myxoid liposarcoma, seems not to be involved in lipomas (Åman et al. 1992). Well-differentiated liposarcomas are cytogenetically very similar to atypical lipomas, i.e., the aberrations are primarily ring and/or giant marker chromosomes (see above). Pleomorphic liposarcomas frequently seem to have very complex karyotypic changes and no abvious similarities with the aberrations in lipomas have been described (Sreekantaiah et al. 1992).

6.2 Lipomas and Other Solid Tumors

Aberrations involving 12q13–15 have been reported in a variety of other tumors, mostly benign ones. These include the cytogenetically well-investigated pleomorphic adenomas of the salivary gland and uterine leiomyomas, in which subgroups have 12q13–15 aberrations, preferentially the t(9;12) and t(12;14), respectively (Bullerdiek et al. 1987; Nilbert and Heim 1990; Sandros et al. 1990; Pandis et al. 1991). Clear cell sarcomas of the tendons and aponeuroses have a characteristic t(12;22) (Bridge et al. 1991; Speleman et al. 1992; Stenman et al. 1992), and a variety of different 12q13–15 aberrations have been reported in chondromas and a subgroup of hemangiopericytomas (Mandahl et al. 1990, 1993a,b; Sreekantaiah et al. 1991a).

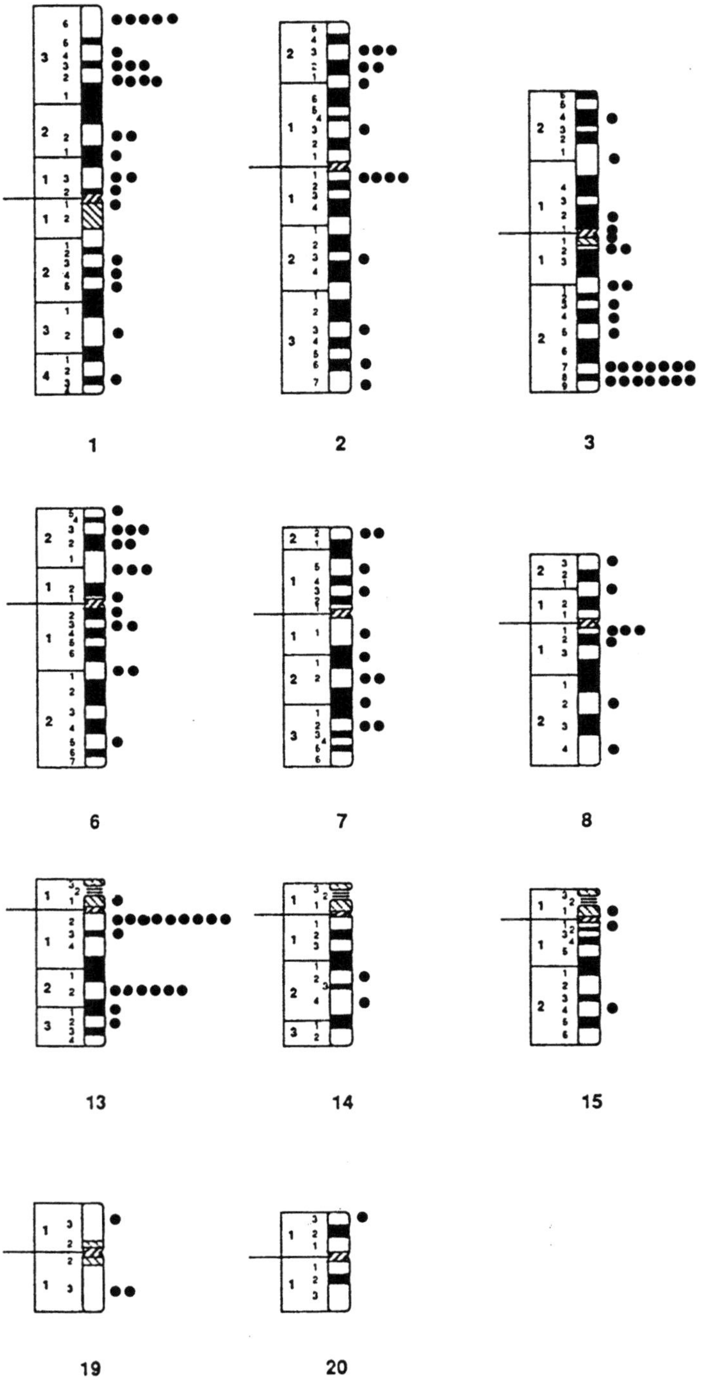

Fig. 4. Breakpoint distribution in all adipose tissue tumors referred to in the text. *Bars* represent 10 breakpoints and *dots* one breakpoint

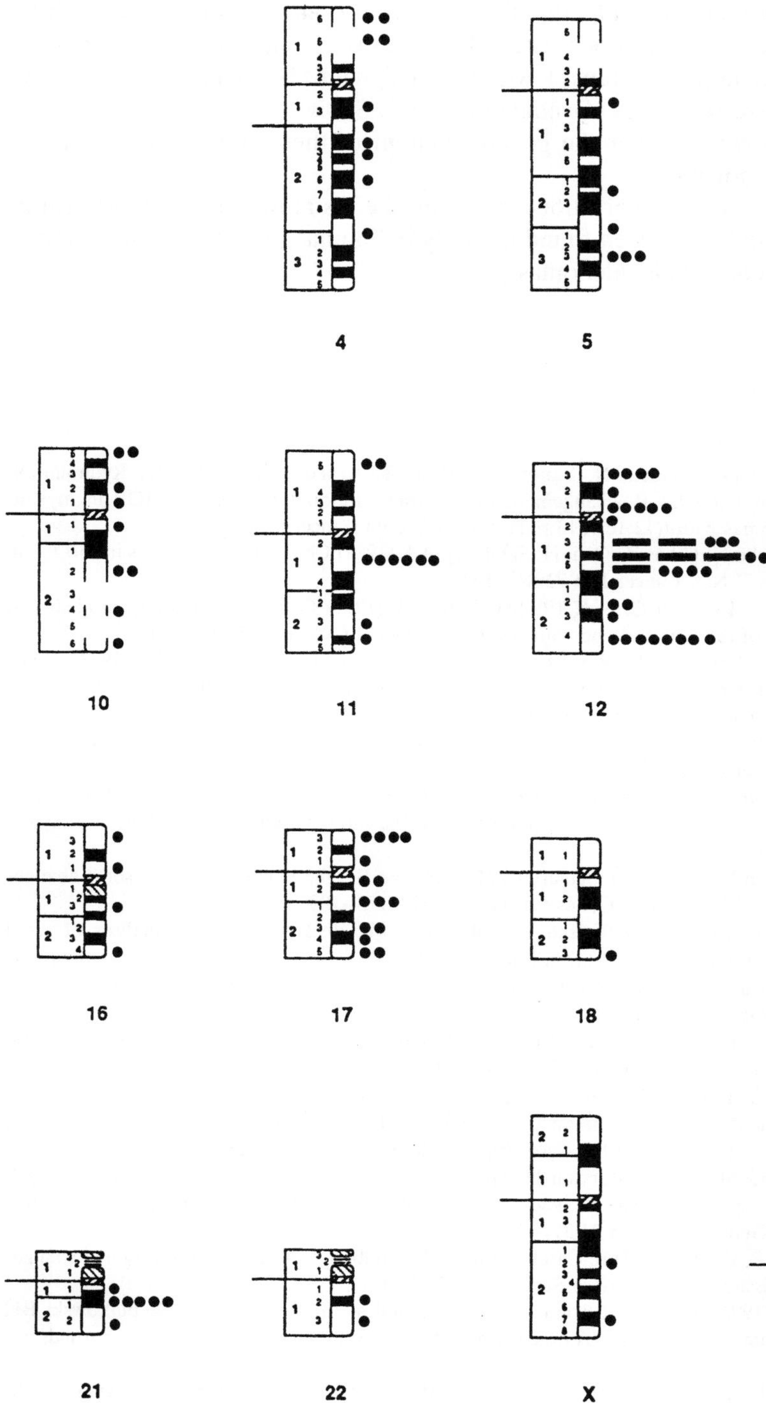

Fig. 4. *Continued*

Supernumerary, unidentifiable ring chromosomes in karyotypes with no or few other aberrations are also consistent aberrations in dermatofibrosarcoma protuberans, low-grade malignant fibrous histiocytomas, and parosteal osteosarcomas (Örndal et al. 1992; Sinovic et al. 1992). Obviously, it is not known whether the genetic content of these rings is similar in the different tumor types.

Finally, various aberrations involving the distal half of the short arm of chromosome 6 have been found not only in lipomas, but also in the leiomyomas and pleomorphic adenomas.

References

Åman P, Ron D, Mandahl N, Fioretos T, Heim S, Arheden K, Willén H, Rydholm A, Mitelman F (1992) Rearrangement of the transcription factor gene CHOP in myxoid liposarcomas with t(12;16)(q13;p11). Genes Chrom Cancer 5:278–285

Becher R, Wake N, Gibas Z, Ochi H, Sandberg AA (1984) Chromosome changes in soft tissue sarcomas. J Natl Cancer Inst 72:823–831

Bridge JA, Sreekantaiah C, Neff JR, Sandberg AA (1991) Cytogenetic findings in clear cell sarcoma of tendons and aponeuroses. Cancer Genet Cytogenet 52:101–106

Bullerdiek J, Bartnitzke S, Weinberg M, Chilla R, Haubrich J, Schloot W (1987) Rearrangements of chromosome region 12q13–q15 in pleomorphic adenomas of the human salivary gland (PSA). Cytogenet Cell Genet 45:187–190

Dal Cin P, Rao U, Turc-Carel C, Sandberg AA (1988a) Translocation (7;21) in a lipoma. Cancer Genet Cytogenet 30:17–22

Dal Cin P, Turc-Carel C, Sandberg AA (1988b) Consistent involvement of band 12q14 in two different translocations in three lipomas from the same patient. Cancer Genet Cytogenet 31:237–240

Dal Cin P, Van Damme B, Hoogmartens M, Van Den Berghe H (1992) Chromosome changes in a case of hibernoma. Genes Chrom Cancer 5:178–180

Debiec-Rychter M, Saryusz-Wolska H, Salagierski M (1992) Cytogenetic analysis of renal angiomyolipoma. Genes Chrom Cancer 4:101–103

De Jong B, Castedo SMMJ, Oosterhuis JW, Dam A (1988) Trisomy 7 in a case of angiomyolipoma. Cancer Genet Cytogenet 34:219–222

Haus O, Collin F, Fraisse J, Mark-Vendel E, Mugneret F, Turc-Carel C (1990) New example of 6p involvement in lipoma. Cancer Genet Cytogenet 48:281–283

Heim S, Mitelman F (1987) Cancer cytogenetics. Liss, New York

Heim S, Mandahl N, Kristoffersson U, Mitelman F, Rööser B, Rydholm A, Willén H (1986) Reciprocal translocation t(3;12)(q27;q13) in lipoma. Cancer Genet Cytogenet 23:301–304

Heim S, Mandahl N, Kristoffersson U, Mitelman F, Rööser B, Rydholm A, Willén H (1987) Marker ring chromosome – a new cytogenetic abnormality characterizing lipogenic tumors? Cancer Genet Cytogenet 24:319–326

Karakousis CP, Dal Cin P, Turc-Carel C, Limon J, Sandberg AA (1987) Chromosomal changes in soft tissue sarcomas. Arch Surg 122:1257–1260

Mandahl N (1992) Methods in solid tumor cytogenetics. In: Rooney DE, Czepulkowski BH (eds) Human cytogenetics. A practical approach: 2. Malignancy and acquired abnormalities. IRL Press, Oxford

Mandahl N, Heim S, Johansson B, Bennet K, Mertens F, Olsson G, Rööser B, Rydholm A, Willén H, Mitelman F (1987) Lipomas have characteristic structural chromosomal rearrangements of 12q13–14. Int J Cancer 39:685–688

Mandahl N, Heim S, Arheden K, Rydholm A, Willén H, Mitelman F (1988) Three major cytogenetic subgroups can be identified among chromosomally abnormal solitary lipomas. Hum Genet 79:203–208
Mandahl N, Heim S, Arheden K, Rydholm A, Willén H, Mitelman F (1990) Chromosomal rearrangements in chondromatous tumors. Cancer 65:242–248
Mandahl N, Örndal C, Heim S, Willén H, Rydholm A, Bauer HCF, Mitelman F (1993a) Aberrations of chromosome segment 12q13–15 characterize a subgroup of hemangiopericytomas. Cancer 71:3009–3013
Mandahl N, Willén H, Rydholm A, Heim S, Mitelman F (1993b) Rearrangement of band q13 on both chromosomes 12 in a periosteal chondroma. Genes Chrom Cancer 6:121–123
Mitchell ELD, Kelsey A, Armstrong G, Dickson A (1992) Ring chromosomes in a retroperitoneal lipoma of childhood. Genes Chrom Cancer 4:89–91
Mitelman F (1991) Catalog of chromosome aberrations in cancer, 4th edn. Wiley-Liss, New York
Nilbert M, Heim S (1990) Uterine leiomyoma cytogenetics. Genes Chrom Cancer 2:3–13
Ohjimi Y, Iwasaki H, Kaneko Y, Ishiguro M, Ohgami A, Kikuchi M (1992) A case of lipoblastoma with t(3;8)(q12;q11.2). Cancer Genet Cytogenet 62:103–105
Örndal C, Mandahl N, Rydholm A, Nilbert M, Heim S, Åkerman M, Mitelman F (1990) Chromosomal evolution and tumor progression in a myxoid liposarcoma. Acta Orthop Scand 61:99–105
Örndal C, Mandahl N, Rydholm A, Willén H, Brosjö O, Heim S, Mitelman F (1992) Supernumerary ring chromosomes in five bone and soft tissue tumors of low or borderline malignancy. Cancer Genet Cytogenet 60:170–175
Pandis N, Heim S, Bardi G, Flodérus U-M, Willén H, Mandahl N, Mitelman F (1991) Chromosome analysis of 96 uterine leiomyomas. Cancer Genet Cytogenet 55:11–18
Peulve P, Thomine E, Hemet J (1990) Cytogenetic analysis of a rare case of pediatric myxolipoma. Ann Genet 33:222–224
Sait SNJ, Dal Cin P, Sandberg AA, Leong S, Karakousis C, Rao U, Harris K (1989) Involvement of 6p in benign lipomas. A new cytogenetic entity? Cancer Genet Cytogenet 37:281–283
Sandberg AA (1990) The chromosomes in human cancer and leukemia, 2nd edn. Elsevier, New York
Sandberg AA, Gibas Z, Saren E, Li FP, Limon J, Tebbi CK (1986) Chromosome abnormalities in two benign adipose tumors. Cancer Genet Cytogenet 22:55–61
Sandros J, Stenman G, Mark J (1990) Cytogenetic and molecular observations in human and experimental salivary gland tumors. Cancer Genet Cytogenet 44:153–167
Sinovic JF, Bridge JA, Neff JR (1992) Ring chromosome in parosteal osteosarcoma. Cancer Genet Cytogenet 62:50–52
Speleman F, Colpaert C, Goovaerts G, Leroy JG, Van Marck E (1992) Malignant melanoma of soft parts. Further cytogenetic characterization. Cancer Genet Cytogenet 58:176–179
Sreekantaiah C, Bridge JA, Rao UNM, Neff JR, Sandberg AA (1991a) Clonal chromosomal abnormalities in hemangiopericytoma. Cancer Genet Cytogenet 54:173–181
Sreekantaiah C, Karakousis CP, Leong SPL, Sandberg AA (1991b) Trisomy 8 as a nonrandom secondary change in myxoid liposarcoma. Cancer Genet Cytogenet 51:195–205
Sreekantaiah C, Leong SPL, Karakousis CP, McGee DL, Rappaport WD, Villar HV, Neal D, Fleming S, Wankel A, Herrington PN, Carmona R, Sandberg AA (1991c) Cytogenetic profile of 109 lipomas. Cancer Res 51:422–433
Sreekantaiah C, Karakousis CP, Leong SPL, Sandberg AA (1992) Cytogenetic findings in liposarcoma correlate with histopathologic subtypes. Cancer 69:2484–2495
Stenman G, Kindblom L-G, Angervall L (1992) Reciprocal translocation t(12:22)(q13;q13) in clear-cell sarcoma of tendons and aponeuroses. Genes Chrom Cancer 4:122–127
Stephenson CF, Berger CS, Leong SPL, Davis JR, Sandberg AA (1992) Analysis of a giant marker chromosome in a well-differentiated liposarcoma using cytogenetics and fluorescence in situ hybridization. Cancer Genet Cytogenet 61:134–138

Stern C, Kazmierczak B, Thode B, Rommel B, Bartnitzke S, Dal Cin P, van de Ven W, Van Den Berghe H, Bullerdiek J (1991) Leiomyoma cells with 12q15 aberrations can be transformed in vitro and show a relatively stable karyotype during precrisis period. Cancer Genet Cytogenet 54:223–228

Turc-Carel C, Dal Cin P, Rao U, Karakousis C, Sandberg AA (1986a) Cytogenetic studies of adipose tissue tumors: I. A benign lipoma with reciprocal translocation t(3;12)(q28;q14). Cancer Genet Cytogenet 23:283–289

Turc-Carel C, Limon J, Dal Cin P, Rao U, Karakousis C, Sandberg AA (1986b) Cytogenetic studies of adipose tissue tumors: II. Recurrent reciprocal translocation t(12;16)(q13;p11) in myxoid liposarcomas. Cancer Genet Cytogenet 23:291–299

Turc-Carel C, Dal Cin P, Boghosian L, Leong SPL, Sandberg AA (1988) Breakpoints in benign lipoma may be at 12q13 or 12q14. Cancer Genet Cytogenet 36:131–135

Involvement of Chromosome 12 in Well-Differentiated Liposarcoma

F. Pedeutour[1], A. Lamouroux[1], J.M. Coindre[2], F. Collin[3], K. Huffermann[1], and C. Turc-Carel[1]

1 Introduction

It is well known that "accurate classification of a given tumor is often not easy, so much that acknowledged international experts are still unable to pin a label on about 10% of sarcomas" (Fletcher 1990). In this context, the characterization of new tumor markers is of paramount interest for the diagnosis, prognosis, and classification of soft tissue tumors (Sandberg and Turc-Carel 1987; Molenaar et al. 1989; Fletcher et al. 1991).

Next to malignant fibrous histiocytoma, liposarcoma is the most common soft tissue sarcoma of adult life (Enzinger and Weiss 1988). Adipose tissue tumors are the soft tissue tumors which are most extensively studied by cancer cytogeneticists. Cytogenetic data have been accumulated that correlate very well with histopathological forms. Specific reciprocal translocations (or less frequently other chromosomal changes) involving two different regions on chromosome 12 characterize myxoid liposarcoma with the t(12;16)(q13;p11) and lipoma with translocations at 12q13–14 (Sandberg 1990). A more distal breakpoint of the 12q13–15 region has also been found in subsets of tumors of different histological background, such as pleomorphic adenomas of the parotid gland (Bullerdiek et al. 1987; Mark et al. 1988), uterine leiomyomas (Heim et al. 1988; Turc-Carel et al. 1988; Vanni and Lecca 1988), chondromas (Mandahl et al. 1993). In contrast to the benign tumors with specific chromosomal aberrations at 12q13–15, the breakpoint for the malignant liposarcomas is located more proximally. As recently shown by Aman et al. (1992), myxoid liposarcomas have a rearrangement of the CHOP gene that is thought to be an essential step for tumorigenesis of the myxoid liposarcomas. In contrast to myxoid liposarcoma, the highly malignant pleomorphic liposarcoma is distinctively

[1] Laboratoire de Cytogénétique Cancérologique, URA CNRS 1462, Université de Nice – Sophia Antipolis, Faculté de Médicine, Avenue de Valombrose, 06107 Nice Cedex 02, France
[2] Laboratoire d'Anatomie Pathologique, Fondation Bergonié, 180 rue de Saint-Genes, 33076 Bordeaux Cedex, France
[3] Laboratoire d'Anatomie Pathologique, Centre George-Francois Leclerc, rue du Professeur Marion, 21034 Dijon Cedex, France

characterized by very high chromosome number counts (70–200), multiple, complex structural rearrangements, and translocated segments which, on most chromosomes, could not be identified (see references in Sreekantaiah et al. 1992).

Although no individual characteristic chromosome changes, either numerical or structural, have been found in well-differentiated liposarcoma (WDLPS), recurrent cytogenetic profiles distinguish them clearly from the other types of malignant or benign lipomatous tumors. Whereas the chromosome number is usually near-diploid, the most striking cytogenetic features are ring chromosomes (one to five copies), large markers with homogeneously stained regions or abnormal, banded regions known to represent cytologic pictures of DNA sequence amplifications, and random telomeric associations which give rise to apparently multicentric giant markers. These abnormalities can be seen together within a single metaphase, in addition to more classical structural chromosome abnormalities. By contrast, other cases show a very simple karyotype with only supernumerary large marker(s) or ring chromosome(s) in an otherwise normal metaphase (Karakousis et al. 1987; Sreekantaiah et al. 1992; Stephenson et al. 1992). In none of these cytogenetically studied WDLPS were the chromosomes at the origin of the rings or large markers identified, due to the difficulty in ascertaining the nature of complex changes by standard cytogenetic banding techniques. We have used fluorescent in situ hybridization (FISH) techniques with centromeric and whole chromosome painting (WCP) probes to tentatively characterize such markers in two cases of WDLPS as an example.

2 Material and Methods

2.1 Investigated Tumors

The tumor T131 was a large retroperitoneal mass (33 × 28 × 60 cm) surgically excised in November 1991 in a 60-year-old woman. The histology of this primary tumor was WDLPS from the lipoma-like subtype with dedifferentiated areas.

The tumor T132 was a local recurrence of a retroperitoneal mass originally diagnosed to be a WDLPS in April 1989. The recurrence excised in November 1991 (14 × 8 cm in size) was from the sclerosing subtype.

2.2 Standard Cytogenetic Analysis

The surgically removed specimens were processed according to Limon et al. (1986). After mechanical and enzymatic dissociation (collagenase Sigma

200 U/ml; 16 h), the suspended tumor cells were cultured in flasks in RPMI 1640 medium supplemented with 10% fetal calf serum, L-glutamine (200 m*M*), and antibiotics. Chromosomal analyses were carried out on short-term cultures (3–12 days). Cultures were arrested by colcemid (Gibco, 0.01 mg/ml, 3–16 h). Hypotonic treatment with 0.075 *M* KCl, fixation with methanol: acetic acid (3:1), and RHG (R-bands by heating using Giemsa) and GTG (G-bands by trypsin using Giemsa) banded air-dried preparations were performed according to standard procedures.

2.3 Molecular Probes

We used the biotinylated alpha-satellite probes provided by Oncor, USA, specific to the following chromosomes: 1–12 (D12Z3), 15–18, 20, and X. Biotinylated WCP probes specific to chromosomes 1, 2, 3, 4, 6, and 12 were a gift from Imagenetics, USA. Direct-labeled orange WCP DNA probes specific to chromosomes 1–4, 6–9, 12, 13, 15, and 18 were purchased from Gibco-BRL, France.

2.4 Fluorescent In Situ Hybridization Techniques

Hybridization and detection were performed according to the method previously described by Pinkel et al. (1983) with slight modifications. Metaphase spreads from fixed cells were stored in 70% ethanol at 4°C. Slides were air dried, treated with RNase, and denatured by immersion in 70% formamide in 2 × saline sodium citrate at 70°C. Labeled probes were treated according to the manufacturers' recommendations. Slides were examined for visualization of the fluorescent signal using a Zeiss Axiophot microscope.

3 Results

3.1 Standard Cytogenetics

WDLPS cells in culture soon stop growing, giving few metaphases available for analysis. Hence, to allow FISH analysis, only five to ten metaphases were studied by standard cytogenetics. For the T131 tumor, ten metaphases were analyzed. The karyotype was 47,XX,+mar. The only change was a supernumerary abnormally banded giant chromosome (Fig. 1a). After CBG (C-bands by barium hydroxide using Giemsa) banding, the centromere of the giant marker was C-band negative, in contrast to the centromeres of all the other normal chromosomes, which were C-band positive (data not shown).

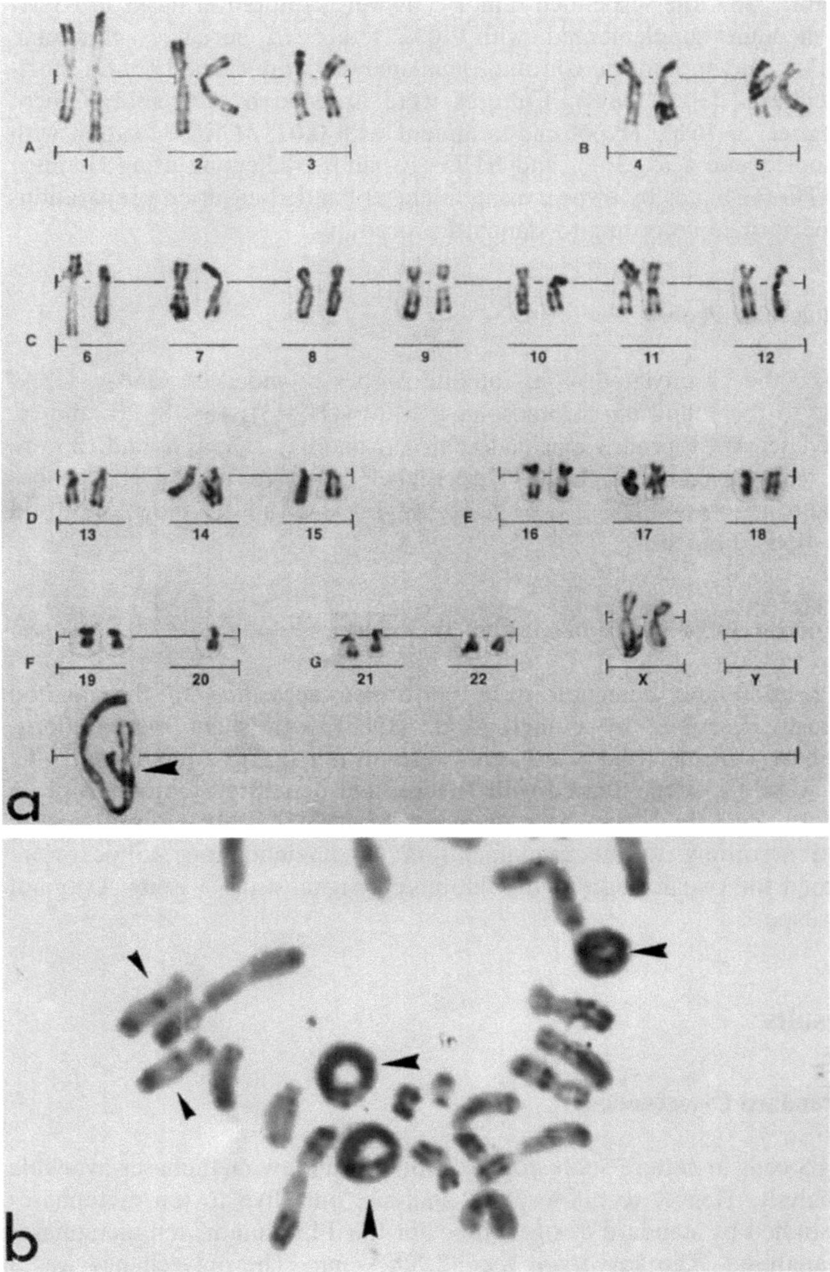

Fig. 1a,b. Characteristic chromosome abnormalities involving chromosome 12 in well-differentiated liposarcoma (WDLPS). Large marker chromosome in a R-banded metaphase with an otherwise normal karyotype from T131 (**a**) and ring chromosomes in a partial R-banded metaphase from T132 (**b**). The *large arrowheads* point to the large marker and rings, the *small arrowheads* to normal chromosome 12s

For the T132 tumor, five R-banded metaphases were analyzed. The karyotype was 47–50,XX, with one to four rings of medium size (Fig. 1b).

3.2 Fluorescent In Situ Hybridization Analysis

In both cases, sequences of chromosome 12 were detected on the supernumerary giant marker and rings using the WCP probe specific to chromosome 12. However, these markers exhibited distinctive fluorescent pattern. In the T131 cells, the signal was distributed in the fashion of a banding pattern over the entire length of the giant marker (Fig. 2a), alternating with regions painted with WCP probes specific to chromosomes 1 and 8 (data not shown). In the T132 cells, the signal with the 12 WCP probe was homogeneously distributed along the entire ring, but with an intensity weaker than that of the two otherwise normal chromosome 12s (Fig. 2b). Since a positive signal was also observed on the rings with the chromosome 16 WCP probe, with a similar weakness to the 12 signal (data not shown), we suggest that material of both chromosomes 12 and 16 are part of the rings, in a very tight intrication.

None of the alpha-satellite probes studied showed positive signals in the centromeric areas of the giant marker of the T131 cells or of the ring chromosomes in the T132 cells.

4 Discussion

In the two WDLPS cases which are reported here, the sole chromosomal abnormalities were a giant marker (T131) and ring chromosomes (1–4) (T132), found in both cases in an otherwise normal karyotype. Both of these types of chromosome markers were consistently painted with the WCP probe specific to chromosome 12 and hence contained chromosome 12 sequences, but none had a positive signal in the centromeric area with the D12Z3 probe, indicating that they do not contain alpha-satellite sequences from chromosome 12. In both markers, additional chromosome material was added to the 12 sequences from the chromosomes 1 and 8 (T131) and 16 (T132), but once again without any positive signal to probes specific to alpha-satellite sequences of these chromosomes.

The first point of interest which may be outlined from these preliminary findings in WDLPS is the consistent involvement of chromosome 12 sequences, which, if confirmed in additional cases, may have some prognostic implications. WDLPS is a malignant adipose tissue tumor with a variable clinical course – which can be very favorable, as is reflected in the term "atypical lipoma," which is used as a synonym of the nonretroperitoneal form of WDLPS (Evans et al. 1979), or which can be

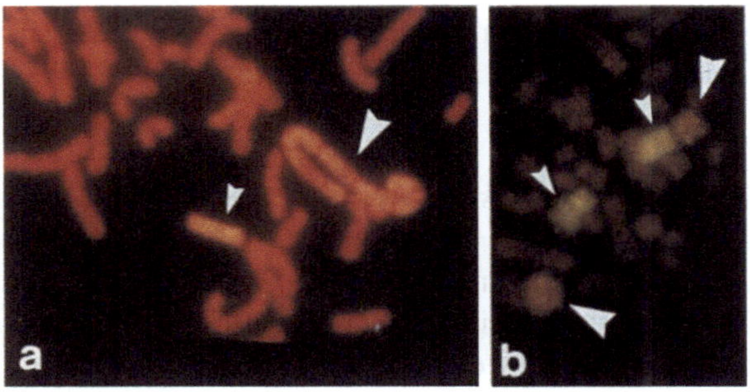

Fig. 2a,b. The large marker and the rings are shown to contain chromosome 12 sequences after fluorescent in situ hybridization (FISH) painting with probes specific to this chromosome (**a** and **b**, respectively). The *large arrowheads* point to the large marker and rings, the *small arrowheads* to normal chromosome 12s

more severe, leading to death due to local invasion when located in the retroperitoneum. By contrast, myxoid liposarcoma has a very distinctive aggressive biological behavior, but shares with WDLPS a specific involvement of chromosome 12. The question which needs to be answered is whether there is an underlying biological significance of the striking involvement of chromosome 12 in distinctive chromosome changes in various distinctive histopathologic forms of lipomatous tumors. It is likely that the answer lies in the nature of the genes in that region and their rearrangement. The CHOP gene is structurally rearranged in myxoid liposarcoma (Aman et al. 1992), although other genes in very close proximity – the MDM2 and SAS genes – are amplified, as recently found in liposarcomas not otherwise histologically and cytogenetically specified (Oliner et al. 1992; Smith et al. 1992). The second point of interest is the mechanism by which the rings and the giant marker occurred, bearing in mind the absence at their centromeric area of any of the alpha-satellite DNA sequences tested, as reported in another giant marker in WDLPS (Stephenson et al. 1992), and the "harlequin" pattern of involvement of sequences from several distinctive chromosomes.

Acknowledgements. This work was supported by CNRS and grants from INSERM, the Caisse Regionale d'Assurances Maladies, the Association pour la Recherche sur le Cancer, and the Federation des Centres de Lutte Contre le Cancer, France.

References

Aman P, Ron D, Mandahl N, Fioretos T, Heim S, Arheden K, Willén H, Rydholm A, Mitelman F (1992) Rearrangement of the transcription factor gene CHOP in myxoid liposarcoma with t(12;16)(q13;p11). Genes Chrom Cancer 5:278–285

Bullerdiek J, Bartnitzke S, Weinberg M, Chilla R, Haubrich J, Schloot W (1987) Rearrangements of chromosome 12q13–q15 in pleomorphic adenomas of the human salivary gland (PSA). Cytogenet Cell Genet 45:187–190

Enzinger FM, Weiss SW (1988) Soft tissue tumors, 2nd edn. Mosby, St Louis

Evans HL, Soule EH, Winkelmann RK (1979) Atypical lipoma, atypical intramuscular lipoma, and well differentiated retroperitoneal liposarcoma. A reappraisal of 30 cases formerly classified as well differentiated liposarcoma. Cancer 43:574–584

Fletcher CDM (1990) Introduction. In: Fletcher CDM, McKee PH (eds) Pathobiology of soft tissue tumors. Churchill Livingstone, Edinburgh, pp 1–7

Fletcher JA, Kozakewich HP, Hoffer FA, Lage JM, Weidner N, Tepper R, Pinkus GS, Morton CC, Corson JM (1991) Diagnostic relevance of clonal cytogenetic aberrations in malignant soft-tissue tumors. N Engl J Med 324:436–442

Heim S, Nilbert M, Vanni R, Floderus U-M, Mandahl N, Liedgren S, Lecca U, Mitelman F (1988) A specific translocation, t(12;14)(q14–q15;q23–q24), characterizes a subgroup of uterine leiomyomas. Cancer Genet Cytogenet 32:13–17

Karakousis CP, Dal Cin P, Turc-Carel C, Limon J, Sandberg AA (1987) Chromosomal changes in soft-tissue sarcomas: a new diagnostic parameter. Arch Surg 122:1257–1260

Limon J, Dal Cin P, Sandberg AA (1986) Application of long term collagenase disaggregation for the cytogenetic analysis of human solid tumors. Cancer Genet Cytogenet 23:305–313

Mandahl N, Willén H, Rydholm A, Heim S, Mitelman F (1993) Rearrangement of band q13 on both chromosomes 12 in a periosteal chondroma. Genes Chrom Cancer 6:121–123

Mark J, Sandros J, Wedell B, Stenman G, Ekedahl C (1988) Significance of the choice of tissue culture technique on the chromosomal patterns in human mixed salivary gland tumors. Cancer Genet Cytogenet 33:229–244

Molenaar WM, De Jong B, Buist J, Idenburg VJS, Seruca R, Vos AM, Hoekstra HJ (1989) Chromosomal analysis and classification of soft tissue sarcomas. Lab Invest 60:266–274

Oliner JD, Kinzler KW, Meltzer PS, George DL, Vogelstein B (1992) Amplification of a gene encoding a p53-associated protein in human sarcomas. Nature 358:80–83

Pinkel D, Straume T, Gray JW (1983) Cytogenetic analysis using quantitative, highly sensitive, fluorescence hybridization. Proc Natl Acad Sci USA 83:2934–2938

Sandberg AA (1990) The chromosomes in human cancer and leukemia, 2nd edn. Elsevier, New York

Sandberg AA, Turc-Carel C (1987) The cytogenetics of solid tumors. Relation to diagnosis, classification and pathology. Cancer 59:387–395

Smith SH, Weiss SW, Jankowski SA, Coccia MA, Meltzer PS (1992) SAS amplification in soft tissue sarcoma. Cancer Res 53:3746–3749

Sreekantaiah C, Karakousis CP, Leong SPL, Sandberg AA (1992) Cytogenetic findings in liposarcoma correlate with histopathologic subtypes. Cancer 69:2484–2495

Stephenson CF, Berger CS, Leong SPL, Davis JR, Sandberg AA (1992) Analysis of a giant marker chromosome in a well-differentiated liposarcoma using cytogenetics and fluorescence in situ hybridization. Cancer Genet Cytogenet 61:134–138

Turc-Carel C, Dal Cin P, Boghosian J, Terk-Zakarian J, Sandberg AA (1988) Consistent breakpoints in region 14q22–q14 in uterine leiomyoma. Cancer Genet Cytogenet 32:25–31

Vanni R, Lecca U (1988) Involvement of the long arm of chromosome 12 in chromosome rearrangement of uterine leiomyoma. Cancer Genet Cytogenet 32:33–34

Breakpoints and Recipient Chromosomes in Pleomorphic Adenomas, Lipomas, and Uterine Leiomyomas – What Do They Tell Us?

M.E. Herrmann[1], G. Belge[2], C. Stern[2], P. Dal Cin[3], J. Bullerdiek[2], and S. Bartnitzke[2]

It has now been well documented that several tumors of epithelial and mesenchymal origin have typical or even specific chromosomal alterations involving 12q13–15. The currently best investigated examples are pleomorphic adenomas (Mark et al. 1980, 1988; Bullerdiek et al. 1987a,b), lipomas (Heim et al. 1986; Turc-Carel et al. 1986a; Mandahl et al. 1988), myxoid liposarcomas (Turc-Carel et al. 1986b), and uterine leiomyomas (Heim et al. 1988; Turc-Carel et al. 1988; Vanni and Lecca 1988). Similar results have recently been obtained from other benign tumors, e.g., cervical polyps (Vanni et al. 1993) or epithelial hyperplasia of the breast (Rohen et al. 1993). For the group of pleomorphic adenomas, lipomas, myxoid liposarcomas, and uterine leiomyomas, Schoenmakers et al. (1993) were able to show that the specific breakpoints in all tumors mentioned above fell within the 7 cM region between loci D12S19 and D12S8. Other chapters of this book have dealt with details of the cytogenetic findings in these tumors. Usually, the aberrations of chromosomal region 12q13–15 are the only karyotypic abnormality, thus indicating that a causal relationship between the cytogenetic changes and tumorigenesis exists. However, up to now the molecular analysis of the mechanisms underlying the chromosomal alterations has remained open and can be expected to fully elucidate as to how these changes contribute to tumorigenesis. Only in one case the molecular basis of a translocation observed is known. This sole exception is the myxoid liposarcoma. Recently, it has been described that the simple t(12;16) or complex types of this translocation found in all of the myxoid liposarcomas analyzed lead to a rearrangement of the CHOP gene which is neither affected by the 12q13–15 changes in pleomorphic adenomas nor in the benign mesenchymal tumors (Aman et al. 1992). Thus, at least a second breakpoint cluster must exist which is more distal to the breaks observed in myxoid liposarcomas and cannot be assigned to a certain gene or certain genes, yet (Mrozek et al. 1993; Aman et al. 1992). Nevertheless, given that the alteration of only one gene is a common event in the benign tumors

[1] Department of Surgery K8, Henry Ford Hospital, Detroit, MI 48202, USA
[2] Centre of Human Genetics and Genetic Counselling University of Bremen 28359 Bremen, Germany
[3] Center of Human Genetics, K.U. Leuven 3000 Leuven, Belgium

mentioned above, it can be concluded from the structural alterations observed that the formation of a fusion gene is less likely than the simple (re)activation of a yet unknown gene since in all tumors variant translocations are more common than the specific ones or, respectively, there is no clearly predominant translocation at all akin to the pleomorphic adenomas. This is a marked contrast to the myxoid liposarcomas in which only the t(12;16) either as a simple reciprocal or a complex translocation has been found. Also the DNA of the benign tumors carrying a 12q13–15 rearrangement is not active in the HIH3T3 assay (own unpublished results).

In summary, we can speculate that the chromosomal rearrangements seen in the benign tumors lead to the activation of a proliferation-associated gene which is weakly inactivated in the normal tissue, because it can easily be switched on by a variety of position effects resulting from the translocations to different partners. However, systematic strategies can be set up to clone the breakpoint regions in the benign tumor and to establish the relationship between the cytogenetic alterations and the underlying changes at the molecular level. On the other hand, akin to the situation in a large number of other tumors characterized by specific translocations as well, very little is known about possible factors inducing these rearrangements. In an attempt to explain the specificity of the primary cytogenetic changes in human tumors Mitelman and Levan (1981) have hypothesized that these primary changes originate from the influence of an agent causing strictly nonrandom patterns of abnormalities. The benign tumors mentioned above offer an excellent chance to get some good ideas about the origin of the translocations for at least three reasons:

- There are at least three benign tumors, each of them with a cytogenetically well established subtype showing the 12q13–15 rearrangements based on a large number of tumors studied.
- The tumors include epithelial as well as mesenchymal tumors with very different histological patterns.
- Usually, there are rather simple karyotypic patterns allowing all aberrations to be identified.

What can now be concluded from the data obtained from the 12q13–15 cases? First, it is apparent that, when summarizing the different breakpoints which have been observed, there is a strong correlation between the number of breakpoints per chromosome and its relative length. A similar correlation can be found when counting the break events per chromosome, with the exception of 3p in lipomas and 14q in uterine leiomyomas. At first glance, this observation fits well with the idea of a nonrandom distribution of the breaks for the variant translocations. On the other hand, it can also be explained by a higher number of potential activators for the putative gene. This latter assumption is confirmed by the very small overlap between the breakpoints of the translocation partners when comparing the situation in the three types of benign tumors. This apparent specificity of the breakpoints depending on the tissue of origin of the tumors may reflect a specific pattern of open chromatin regions which, probably for the same reason, i.e., an activated chromatin configuration, serve both as translocation hotspots and as activators. There is a significant correlation between these breakpoints and light G bands as can be expected if there is indeed a coincidence between open chromatin regions and their ability to be predisposed to become recipient sites or to undergo somatic recombination, respectively. Another point of considerable interest is whether or not the occurrence of translocation breakpoints is influenced by the presence of fragile sites. Generally, it is yet an unresolved problem, whether such coincidence predisposes carriers of fragile sites of rearrangements that lead to malignant tumors.

Sutherland and Simmers (1988) published a statistical analysis of 91 fragile sites and the cancer breakpoints as collected in the HGM8 (Berger et al. 1985). Taking into account that neither cancer breakpoints nor fragile sites are distributed randomly within the human genome, but seem to occur more frequently within the light G bands, they analyzed the coincidence of breakpoints and fragile sites restricted only to light G bands and found no significant association. However, differentiating between common and rare fragile sites, they found a correlation between breakpoints and rare fragile sites.

Another example of searching for the significance of rare fragile sites is a follow-up study of 100 families with segregating rare fragile sites. An increase of malignancies in these families could not be shown in this study (Le Beau and Rowley 1984). Again a different approach was undertaken by Moreno-Laguna et al. (1988). They studied aphidicolin-induced fragile sites in patients with astrocytomas, meningiomas, and a control group and found a highly significant difference of incidence of lesions between the tumor group and the controls at a $p < 10^{-6}$. Evidence for a coincidence between chromosomal rearrangements in acute myelomonocytic leukaemia at band 16q22 and the rare fragile sites FRA 16B has been disproved by the study of Simmers et al. (1987). In situ hybridization revealed that the tumor breakpoint on band 16q22 is not in the immediate proximity of FRA 16B on the molecular level; thus the metallotheionein gene cluster is split at the tumor breakpoint but not at FRA 16B.

We statistically investigated the possible coincidence of recombination sites in the benign tumors and the fragile sites. Like Sutherland and Hecht (1985), using an analysis of the breakpoints listed in the HGM8, we found, based on the HGM11 data (Mc Alpine et al. 1991), a significant association of breakpoints and fragile sites when a random distribution of breakpoints within the genome was assumed. We also found a preferential location of breakpoints on light bands. Taking this obvious preferential occurrence of tumor breakpoints in light G bands into account for statistical analysis, no significant association was found, neither for common nor for rare fragile sites. Therefore, actually, the breakpoints quite obviously do not favor the location of fragile sites. However, further analysis such as gene linkage studies may finally prove whether there is a coincidence between single fragile sites and breakpoints in pleomorphic adenomas of the parotid gland, lipomas, or uterine leiomyomas.

As generally shown for cancer breakpoints, a significantly nonrandom correlation between the breakpoints and the fragile sites has been observed which most likely can be explained by the fact that the fragile sites as well as the breakpoints are predominantly in light G bands. Therefore, when testing only those breakpoints occurring in light G bands, no significant correlation is found. This finding exactly confirms the findings by Sutherland and Simmers (1988) who demonstrated that there is no statistical association between common fragile sites and cancer breakpoints when their preferential distribution on light G bands is taken into account. However, if factors exist determining the probability of a band to become a breakpoint, it should be worthwhile to test not only the breakpoints, but also the number of individual breaks at each breakpoint. Again, no significant correlation between the breaks and the location of fragile sites was found, indicating that as a rule the breaks in these tumors occur independently of fragile sites.

Although the fact that Giemsa-light bands are the preferential target of breakpoints as well as of break events offers strong evidence that an open chromatin configuration increases its probability to participate in a chromosomal rearrangement in these tumors, the causative agent(s) of these rearrangements remain to be elucidated. Two different models can be supported to explain these phenomena. The first alternative is a stochastic model in which the mitotic recombination underlying the chromosomal rearrangement is a result of previous more or less random DNA damage due, for example, to X-rays followed by selection for particular rearrangements. In that case a preferential clustering to light G bands seems to be very likely since it is well known that "open" chromatin regions are more sensitive to DNA damage. The other model is an active model in which the rerrangement is due to mitotic recombination within intact DNA strands. In contrast to the former model there may be factors which, by stimulating cellular recombinases, directly lead to recombination of particular regions of open chromatin. This model also fits well with the observation of a preferred location of breaks in Giemsa-light bands since "open" chromatin is known to be more recombigenic. Neither of the two assumptions can presently be validated. Nevertheless, the fact that structural rearrangements involving one break, i.e., simple terminal deletions, are very rare events even among karyotypic changes involving large numbers of breaks speaks in favor of the second model, in which the specificity of the aberrations observed is not due to a selection of single cells having randomly distributed breakpoints but of particular rearrangements alone.

In summary, although it is without any doubt a major challenge to clone the breakpoints and to further elucidate the changes resulting from the specific rearrangements, the question of how these changes originate should by no means be underestimated.

Acknowledgement. This work was supported in part by a grant to MEH by the Surgical Clinic Charlottenburg, Free University Berlin.

References

Aman P, Ron D, Mandahl N, Fioretos T, Heim S, Arheden K, Willen H, Rydholm A, Mitelman F (1992) Rearrangement of the transcription factor gene CHOP in myxoid liposarcoma with t(12;16)(q13;p11). Genes Chrom Cancer 5:278–285

Berger R, Bloomfield CD, Sutherland GR (1985) Report of the committee on chromosome rearrangements in neoplasia and on fragile sites (HGM8). Cytogenet Cell Genet 40:516–520

Bullerdiek J, Böschen C, Bartnitzke S (1987a) Aberrations of chromosome 8 in mixed salivary gland tumors - cytogenetic findings on seven cases. Cancer Genet Cytogenet 24:205–212

Bullerdiek J, Raabe G, Bartnitzke S, Böschen C, Schloot W (1987b) Structural rearrangements of chromosome 8 involving 8q12 – a primary event in pleomorphic adenoma of the parotid gland. Genetica 72:85–92

Heim S, Mandahl N, Kristoffersson U, Mitelman F, Röser B, Rydholm A, Willen H (1986) Reciprocal translocation t(3;12)(q27;q13) in lipoma. Cancer Genet Cytogenet 23:301–304

Heim S, Nilbert M, Vanni R, Floderus UM, Mandahl N, Liedgren S, Lecca U, Mitelman F (1988) A specific translocation, t(12;14)(q14–q15;q23–q24), characterizes a subgroup of uterine leiomyomas. Cancer Genet Cytogenet 32:13–17

Le Beau MM, Rowley JD (1984) Heritable fragile sites in cancer. Nature 308:607–608

Mandahl N, Heim S, Arheden K, Rydholm A, Willen H, Mitelman F (1988) Three major cytogenetic subgroups can be identified among chromosomally abnormal solitary lipomas. Hum Genet 79:203–208

Mark J, Dahlenfors R, Ekedahl C, Stenman G (1980) The mixed salivary gland tumor: a normally benign human neoplasm frequently showing specific chromosomal abnormalities. Cancer Genet Cytogenet 2:231–241

Mark J, Sandros J, Wedell B, Stenman G, Ekedahl C (1988) Significance of the choice of tissue culture technique on the chromosomal patterns in human mixed salivary gland tumors. Cancer Genet Cytogenet 33:229–244

Mc Alpine PJ, Shows TB, Boucheix C, Huebner M, Anderson WA (1991) The 1991 catalog of mapped genes and report of the nomenclature committee (HGM11). Cytogenet Cell Genet 58:5–102

Mitelman F, Levan G (1981) Clustering of aberrations to specific chromosomes in human neoplasms. IV. A survey of 1871 cases. Hereditas 95:79–139

Moreno-Laguna S, Campos JM, Kusack ME, Bello MJ, Ramos-Arroyo MA, Rey JA (1988) Common fragile sites in meningioma, astrocytoma, and control patients. Book of abstracts. First European workshop on cytogenetics and molecular genetics of human solid tumors, Dijon

Mrozek K, Karakousis CP, Bloomfield CD (1993) Chromosome 12 breakpoints are cytogenetically different in benign and malignant lipogenic tumors: localization of breakpoints in lipoma to 12q15 and in myxoid liposarcoma to 12q13.3. Cancer Res 53:1670–1675

Rohen C, Bonk U, Staats B, Bartnitzke S, Bullerdiek J (1993) Two human breast tumors with translocations involving 12q13–15 as the sole cytogenetic abnormality. Cancer Genet Cytogenet 69:68–71

Schoenmakers EFPM, Kools PFJ, Kazmierczak B, Bartnitzke S, Bullerdiek J, Dal Cin P, Van den Berghe H, Van de Ven WJM (1993) Recurrent chromosome 12 aberrations of four types of solid tumors are clustered between D12S19 and D12S8. Genomics 20:210–222

Simmers RN, Sutherland GR, West A, Richards RI (1987) Fragile sites at 16q22 are not at the breakpoint of the chromosomal rearrangement in AMMol. Science 236:92–94

Sutherland GR, Hecht F (1985) Fragile sites on human chromosomes. Oxford University Press, Oxford

Sutherland GR, Simmers RN (1988) No statistical association between common fragile sites and nonrandom chromosome breakpoints in cancer cells. Cancer Genet Cytogenet 31:9–15

Turc-Carel C, Dal-Cin P, Karakousis C, Sandberg AA (1986a) Cytogenetic studies of adipose tissue tumors. I. A benign lipoma with reciprocal translocation t(3;12)(q28;q14). Cancer Genet Cytogenet 23:283–289

Turc-Carel C, Limon J, Dal Cin P, Rao U, Karakousis C, Sandberg AA (1986b) Cytogenetic studies of adipose tissue tumors. II. Recurrent reciprocal translocation t(12;16)(q13;p11) in myxoid liposarcomas. Cancer Genet Cytogenet 23:291–299

Turc-Carel C, Dal-Cin P, Boghosian J, Terk-Zakarian J, Sandberg AA (1988) Consistent breakpoints in region 14q22–q24 in uterine leiomyoma. Cancer Genet Cytogenet 32:25–31

Vanni R, Lecca U (1988) Involvement of the long arm of chromosome 12 in chromosome rearrangement of uterine leiomyoma. Cancer Genet Cytogenet 32:33–34

Vanni R, Marras S, Moerman P, Andria M, Valdes E, Deprest H, Van den Berghe H, Dal Cin P (1993) Endometrial polyp: another benign tumor characterized by 12q14–15 changes? Abstracts of the 5th International Workshop on Chromosomes in Solid Tumors. Tucson, Arizona, 10–12 January

Trisomy 12 Clusters to Tumors of the Female Genital Tract

M. Kiechle-Schwarz[1], T. Bauknecht[1], and A. Pfleiderer[1]

Trisomy 12 is a very common cytogenetic finding in chronic B-lymphocyte leukemias, frequently occurring as the only abnormality and representing a primary change in these tumors [1]. It has also been sporadically described in some non-Hodgkin's lymphomas [2]. In nonlymphatic solid tumors, however, an extra chromosome 12, especially as the only cytogenetic deviation, has been rarely observed. During our cytogenetic studies on solid tumors of the female genital tract we made the observation that trisomy 12 is frequently found in these tumors. The karyotype of a recently analyzed cystadenofibroma of the ovary is shown in Fig. 1. Moreover, trisomy 12 appears to be a feature almost restricted to neoplasms originating from the female genital organs.

Hitherto isolated trisomies 12 have been reported in 35 solid tumors. Interestingly, 33 of these neoplasms originated from the uterus and the ovaries (Table 1). Trisomy 12 has been reported as a specific, recurrent aberration in leiomyomas [3–6,17], and has also been observed in a malignant uterine tumor (endometrial adenocarcinoma) [7]. Furthermore, it has been found in 21 ovarian tumors, such as fibromas, thecomas, fibrothecomas, and cystadenomas [8–10,14,15,17], two ovarian borderline tumors [8,14], two well and moderately differentiated ovarian carcinomas [8,16], and a malignant teratoma [11]. The only two solid tumors with an extra chromosome 11 not deriving from female genitals, were an adenosquamous carcinoma of the lung [12] and a Wilms' tumor [13]. Interestingly, both tumors occurred in females.

Regarding the gynecological tumors with an additional chromosome 12, 12 tumors were located in the uterus and 21 derived from the ovaries. Of 21 ovarian tumors with trisomy 12, 16 were of benign phenotype: 8 fibromas, 3 fibrothecomas, 2 thecomas, 1 juvenile granulosa cell tumor, and 2 cystadenomas. Two tumors, and adult granulosa cell tumor and an epithelial ovarian tumor, have been of borderline malignancy. Three additional tumors, an immature teratoma and two ovarian carcinomas (grade 1 and 2), were of malignant nature [8,11,16]. Regarding the uterine tumors with trisomy 12, there were mainly 11 totally benign leiomyomas. Among the

[1] Universitäts-Frauenklinik, Hugstetterstrasse 55, 79106 Freiburg, Germany

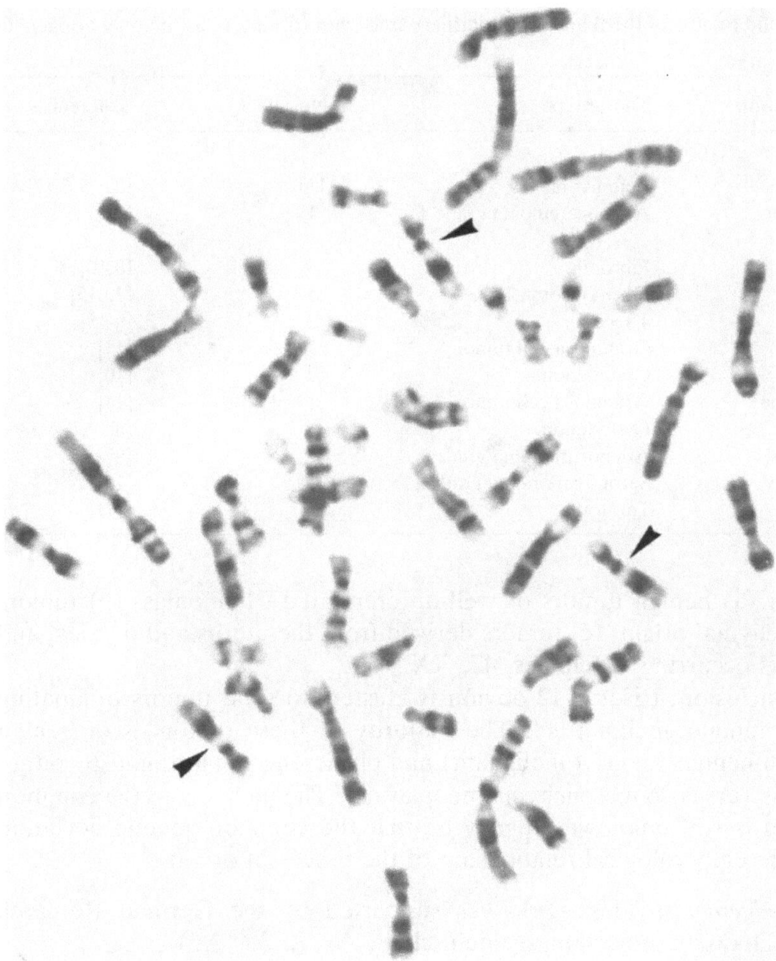

Fig. 1. Metaphase spread of an cystadenofibroma of the ovary showing trisomy 12 in 22 analyzed cells. *Arrows* indicate three chromosomes 12

uterine tumors there was only one malignant tumor, a well-differentiated adenocarcinoma, with an extra chromosome 12. Summarizing the malignant potential of the gynecological tumors with trisomy 12, there were 27 benign tumors, 2 borderline cases, and only 4 malignant tumors. Regarding the histogenesis of these tumors, the vast majority (26 tumors) was of mesenchymal origin, whereas 6 derived from epithelial tissues. Thus, trisomy 12 obviously characterizes a group of solid tumors arising from the uterus and the ovaries which is of benign phenotype and of mesenchymal origin. In addition, all solid tumors with an extra chromosome 12 as the only aberration so far reported occurred in females. The features of these are the

Table 1. Solid tumors of the female genitourinary tract with trisomy 12 as the only cytogenetic deviation

Tumor location	Tumor type	No. of cases	References
Uterus			
Benign	Leiomyoma	11	[3,4,5,6,17]
Malignant	Adenocarcinoma grade 1	1	[7]
Ovary			
Benign	Fibroma	8	[8,10,14,15,17]
	Fibrothecoma	3	[10,15]
	Thecoma	2	[9,15]
	Granulosa cell tumor	1	[15]
	Cystadenoma	2	[10]
Borderline	Granulosa cell tumor	1	[14]
	Cystadenoma	1	[8]
Malignant	Adenocarcinoma grade 1	1	[8]
	Serous carcinoma grade 2	1	[16]
	Teratoma	1	[11]

following: (a) benign tumors or well-differentiated carcinomas, (b) tumors of mesenchymal origin, (c) tumors derived from the uterus and ovaries, and (d) tumors occurring in females, 47, XX, + 12.

In conclusion, trisomy 12 obviously clusters to solid tumors originating from the female genital tract. The majority of these tumors is of benign nature (26 benign versus 4 malignant) and of mesenchymal origin (6 epithelial tumors versus 26 mesenchymal neoplasms). The meaning of the common aberration is still unknown. It may be that the common genetic deviation reflects the embryological relationship of the tissues of origin.

Acknowledgements. This work was supported by the German Research Council (Deutsche Forschungsgemeinschaft, AZ Ki 352-2/2)

References

1. Nowell PC, Moreau L,Growney P, Besa EC (1988) Karyotypic stability in chronic B-cell leukemia. Cancer Genet Cytogenet 33:155–160
2. Mitelman F (1988) Catalog of chromosome aberrations in cancer, 3rd edn. Liss, New York
3. Nilbert M, Heim S, Mandahl N, Floderus U-M, Willen H, Mitelman F (1990) Trisomy 12 in uterine leiomyomas. Cancer Genet Cytogenet 45:63–66
4. Vanni R, Nieddu M, Paoli R, Lecca U (1989) Uterine leiomyoma cytogenetics. I. Rearrangements of chromosome no. 12. Cancer Genet Cytogenet 37:49–54
5. Havel G, Dahlenfors R, Wedell B, Mark J (1989) Similar chromosomal evolution in a uterine stromomyosarcoma and in one of two leiomyomas from the same patient. APMIS 97:143–146
6. Kiechle-Schwarz M, Sreekantaiah C, Berger CS, Pedron S, Medchill MT, Surti U, Sandberg AA (1990) Nonrandom cytogenetic changes in leiomyomas of the female genitourinary tract: a report of 35 cases. Cancer Genet Cytogenet

7. Couturier J, Vielh P, Salmon R, Dutrillaux B (1986) Trisomy and tetrasomy for the long arm of chromosome 1 in near-diploid human endometrial adenocarcinomas. Int J Cancer 41:202-217
8. Samuelson J, Katz S, Schwatz, Yang-Feng TI (1988) Chromosomal evolution through tumor progression in ovarian cancer. Am J Hum Genet [Suppl] A33
9. Mrozek K, Nedoszytko B, Babinska M, Mrozek E, Hrabowska M, Emerich J, Limon J (1990) Trisomy of chromsome 12 in a case of thecoma of the ovary. Gynecol Oncol 36:413–416
10. Pejovic T, Heim S, Mandahl N, Elmfors B, Floderus U-M, Furgyik S, Helm G, Willen H, Mitelman F (1990) Trisomy 12 is a consistent chromosomal aberration in benign ovarian tumors. Genes Chrom Cancer 2:48–52
11. Kusyk CJ, Turpening EL, Edwards CL, Wharton JT, Copeland LJ (1982) Karyotypic analysis of four solid gynecologic tumors. Gynecol Oncol 14:324–338
12. Liang JC, Kurzrock R, Gutterman JU, Gallick GE (1986) Trisomy 12 correlates with elevated expression of p21ras in a human adenosquamous carcinoma of the lung. Cancer Genet Cytogenet 23:183–188
13. Dao DD, Schroeder WT, Chao L-Y, Kikuchi H, Strong LC, Ricardi VM, Pathak S, Nichols WW, Lewis WH, Saunders GF (1987) Genetic mechanisms of tumor-specific loss of 11p DNA sequences in Wilms' tumor. Am J Hum Genet 41:202–217
14. Leung W-Y, Schwartz PE, Ng H-T, Yang-Feng TL (1990) Trisomy 12 in benign fibroma and granulosa cell tumor of the ovary. Gynecol Oncol 38:28–31
15. Fletcher JA, Gibas Z, Donovan K, Perez-Atayde A, Genest D, Morton CC, Lage JM (1991) Ovarian ganulosa-stromal cell tumors are characterized by trisomy 12. Am J Pathol 138:515–520
16. Kiechle-Schwarz M, Schmidt J, Bauknecht T, Walz L, Pfleiderer A (1994) Recurrent chromosome changes in human ovarian cancer. Cancer Detec Prevent (in press)
17. Kiechle-Schwarz M, Sreekantaiah C, Berger CS, Pfleiderer A, Medchill MT, Avery A, Sandberg (1991) Cluster of trisomy 12 to tumors of the female genitourinary tract. Cancer Genet Cytogenet 54:273–275

Numerical Chromosome 12 Analysis of Carcinoma In Situ and Invasive Germ Cell Tumors of the Adult Testis:

A Study Based on the Simultaneous Application of Interphase Cytogenetics and Immunohistochemistry on Tissue Sections

L.H.J. Looijenga, A.J.M. Gillis, W.L.J. Van Putten, and J.W. Oosterhuis

1 Introduction

Limited data are available about the chromosomal constitution of the precursor to all testicular germ cell tumors of human adults (TGCTs) (Walt et al. 1989; Giwercman et al. 1990; Vos et al. 1990) – the so-called carcinoma in situ (CIS) (Skakkebæk et al. 1987). Therefore questions about the early chromosomal events in the pathogenesis of TGCTs still remain to be answered. In addition, possible chromosomal heterogeneity of CIS, in itself and in relation to the histology of invasive tumors – seminomas (SE) and nonseminomatous TGCTs (NS) – is unknown. The chromosomal basis of the heterogeneity found in some invasive TGCTs by ploidy analysis (Oosterhuis et al. 1989; Fosså et al. 1991a,b; Ky Sela and Matoska 1991) is also not understood.

To study these problems in more detail we applied immunohistochemistry (IHC) in combination with nonradioactive in situ hybridization (ISH) using chromosome-specific repetitive DNA probes on frozen tissue sections of 20 μm thickness. Because of the frequent involvement of chromosome 12 in numerical and structural abnormalities in TGCTs (reviewed in De Jong et al. 1990), we focus on these results in this chapter. The answers to these questions might provide new clues about the pathogenetic relationship(s) between CIS, SE, and NS.

2 Materials and Methods

The numeric distribution pattern of chromosome 12 centromeric regions (identified by pα12H8) (Looijenga et al. 1990) of six SE and five NS was

---

Laboratory of Experimental Patho-Oncology, Dr. Daniel den Hoed Cancer Center, Groene Hilledijk 301, 3075 AE Rotterdam, The Netherlands studied, as well their CIS within the same testis. To identify CIS and SE, placental-like alkaline phosphatase (PLAP) expression was used as immunohistochemical marker (Jacobsen and Petersen 1984; Manivel et al. 1987), while cytokeratic 8 and 18 expression (recognized by monoclonal antibody CAM 5.2) was used for the identification of NS cells (Looijenga et al. 1991, and references cited therein).

IHC in combination with ISH was performed as described (Looijenga et al. 1993). Counting was performed by one person, while results were checked at random by another. Proper hybridization was verified by counting the number of signals in nontumor cells. Analysis of the tumor cells was performed only if at least 95% of 300 nuclei of the nontumor cells contained the number of expected signals: two for diploid and one for haploid cells.

Heterogeneity of signal distributions per nucleus was analyzed separately for each specimen, using the chi-square test of independence. The signal distributions per specimen were summarized by the mean number of spots/nucleus (MNSN) and the associated standard deviation (SDSN). Differences between the MNSN of CIS adjacent to SE (CS) and CIS adjacent to NS (CN), SE and NS were analyzed using the paired and unpaired student's t test.

Single-parameter flow cytometry (SPFCM) was performed as described to identify the ploidy of the CIS and tumor cells (Oosterhuis et al. 1989).

3 Results and Discussion

ISH, also known as interphase cytogenetics (Cremer et al. 1986), makes analysis of numerical and structural aberrations in interphase nuclei possible. This technique, in combination with classical IHC on tissue sections, allows interpretation of the chromosomal constitution of specific subpopulations of cells within their histological context. We used this approach to analyze the chromosome 12 centromeric region constitution of invasive TGCTs and their CIS. The DNA content of the different tumor components was studied using SPFCM (Table 1), illustrating aneuploidy.

Concerning the chromosome 12 centromeric regions no localization-dependent distribution pattern was found between different CIS-containing seminiferous tubules within one specimen or between different localizations within one tumor (not shown). Therefore, the MNSN within one specimen is taken as representative for that particular specimen. These data are summarized in Table 2 and are schematically illustrated in Fig. 1. No difference was detected in MNSN of chromosome 12 between CS and SE. In contrast, CN displayed a significantly lower MNSN of chromosome 12 compared to NS ($p < 0.006$). A significantly lower MNSN of chromosome 12 centromeric regions in CN compared to CS was found ($p < 0.014$).

Table 1. DNA index (DI) of seminomatous (SE) and nonseminomatous testicular germ cell tumors (NS) and their adjacent carcinoma in situ (CS and CN, respectively) analyzed by single-parameter DNA flow cytometry using separately processed components

Specimen	DI	Specimen	DI	Specimen	DI	Specimen	DI
CS1	1.62	SE1	1.62	CN1	1.55	NS1	1.43
CS2	1.73	SE2	1.50	CN2	1.60	NS2	1.50
CS3	1.52	SE3	1.55	CN3	1.57	NS3	1.46
CS4	1.51	SE4	1.46	CN4	1.51	NS4	1.57
CS5	1.55	SE5	1.55	CN5	1.63	NS5	1.41
CS6	1.63	SE6	1.60				

Table 2. Summary of the numerical analysis of centromeric regions of chromosome 12 in six seminomas (SE), five nonseminomatous testicular germ cell tumors (NS), and their adjacent carcinoma in situ (CS and CN, respectively): number of locations or seminiferous tubules (n^{loc}), cells (n^{cells}), mean number of spots per interphase nucleus (MNSN), and standard deviations (SDNS)

Case	n^{loc}	n^{cells}	MNSN	SDNS	Case	n^{loc}	n^{cells}	MNSN	SDSN
CS1	11	1294	3.3	0.30	CN1	14	1107	2.7	0.41
CS2	9	396	3.7	0.50	CN2	10	441	1.9	0.20
CS3	12	939	2.6	0.27	CN3	6	465	2.6	0.26
CS4	14	1133	2.6	0.45	CN4	13	796	2.5	0.28
CS5	10	639	3.5	0.81	CN5	10	1049	2.6	0.32
CS6	10	1435	3.5	0.28	Total	53	3858	2.5	0.32
Total	66	5834	3.2	0.47					
SE1	1	982	3.3	0.26	NS1	1	487	4.2	0.48
SE2	2	338	3.3	0.30	NS2	1	361	2.3	0.30
SE3	1	1354	2.7	0.22	NS3	3	702	4.0	0.28
SE4	1	1259	3.3	0.34	NS4	1	660	3.9	0.55
SE5	1	501	3.5	0.40	NS5	1	209	3.6	0.50
SE6	1	1219	3.5	0.25	Total	7	2419	3.5	0.76
Total	7	5653	3.5	0.68					

Our results on the chromosomal constitution of CIS and the adjacent tumor illustrate their heterogeneity, the similarity between CIS and SE, and differences between CIS adjacent to SE and NS. This strongly suggests that CIS undergo partially the same kind of evolution as the invasive tumor component, i.e., net loss of chromosomes after polyploidization (summarized in De Jong et al. 1990). Based on the data presented here, the hypothesis that

Fig. 1. Schematic representation of the distributions (percentages) of centromeric regions of chromosome 12 in the nuclei of tumor cells of six seminomas (SE), five nonseminomatous testicular germ cell tumors (NS), and their adjacent carcinoma in situ (CS and CN)

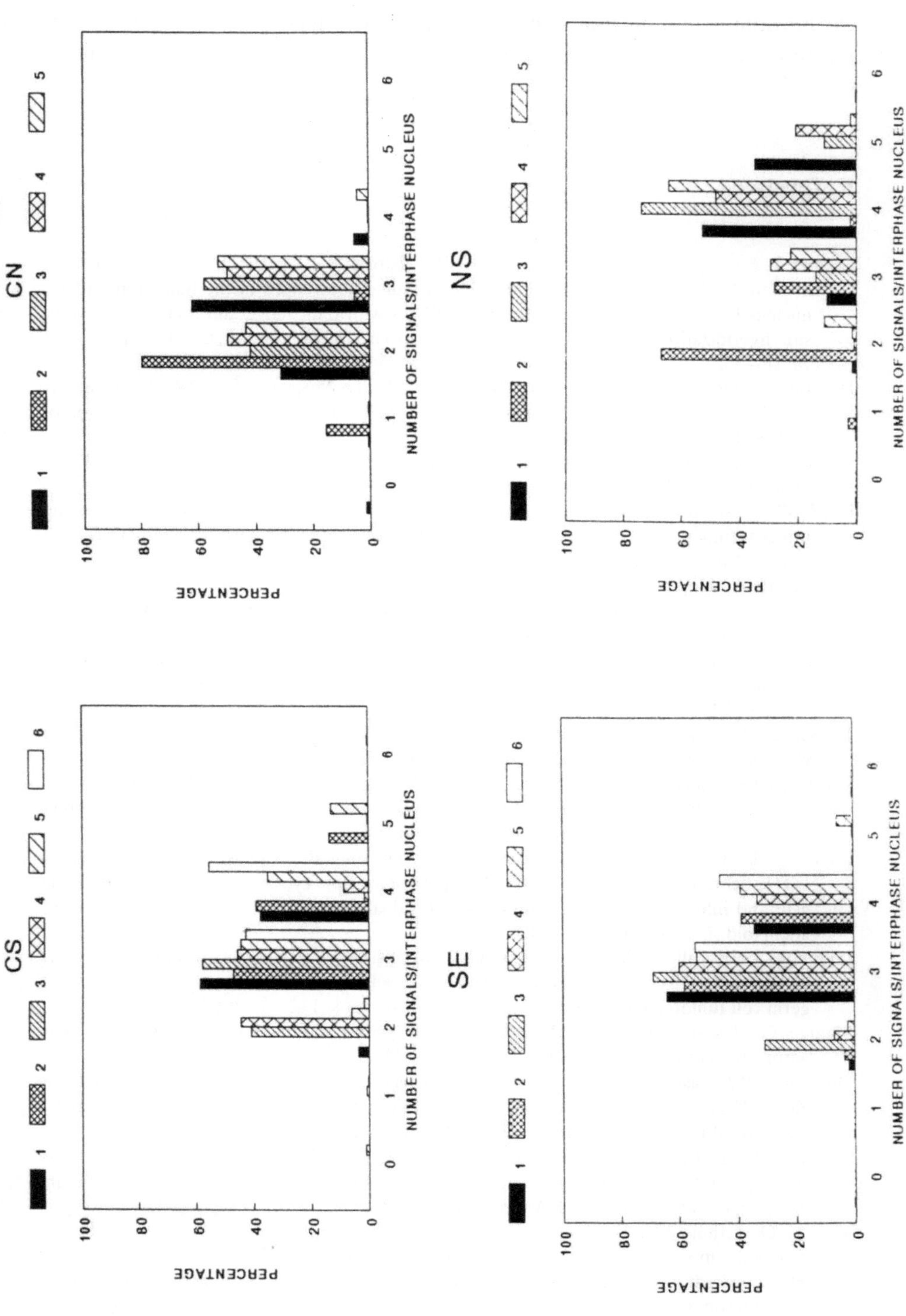
CN
NS
CS
SE
PERCENTAGE
NUMBER OF SIGNALS/INTERPHASE NUCLEUS
0 20 40 60 80 100
0 1 2 3 4 5 6

chromosomal and/or molecular aberrations detected in CIS localized adjacent to an invasive TGCT are by definition early events in the pathogenesis of TGCTs must be reconsidered.

References

Cremer T, Landegent J, Bruckner A, Scholl HP, Schardin M, Hager HD, Devilee P, Pearson P, Van der Ploeg (1986) Detection of chromosome aberrations in the human interphase nucleus by visualization of specific target DNAs with radioactive and non-radioactive in situ hybridization techniques: diagnosis of trisomi 18 with probe L1.84. Hum Genet 74:346–352

De Jong B, Oosterhuis JW, Castedo SMMJ, Vos A, Te Meerman G (1990) Pathogenesis of adult testicular germ cell tumors: a cytogenetic model. Cancer Genet Cytogenet 48:143–167

Fosså SD, Nesland JM, Pettersen EO, Amellem O, Waehre H, Heimdal K (1991a) DNA ploidy in primary testicular cancer. Br J Cancer 64:948–952

Fosså SD, Nesland JM, Waehre H, Amellem O, Pettersen EO (1991b) DNA ploidy in the primary tumor from patients with nonseminomatous testicular germ cell tumors clinical stage I. Cancer 67:1874–1877

Giwercman A, Hopman AGH, Ramaekers FCS, Skakkebæk NE (1990) Carcinoma in situ of the testis. Detection of malignant germ cells in seminal fluid by means of in situ hybridization. Am J Pathol 136:497–502

Jacobsen GK, Pedersen BN (1984) Placental alkaline phosphatase in testicular germ cell tumours and in carcinoma-in-situ of the testis. Acta Pathol Microbiol Immunol Scand [A] 92:323–329

Ky Sela B, Matoska J (1991) Flow cytometric analysis of ploidy and proliferation activity in classical and spermatocytic seminoma. Neoplasma 38:3–11

Looijenga LHJ, Smit VTHBM, Wessels JW, Mollevanger P, Oosterhuis JW, Cornelisse CJ, Devilee P (1990) Localization and polymorphism of a chromosome 12-specific α satellite DNA sequence. Cytogenet Cell Genet 53:216–218

Looijenga LHJ, Oosterhuis JW, Ramaekers FCS, De Jong B, Dam A, Beck JLM, Sleijfer DTH, Schraffordt Koops H (1991) Dual parameter flow cytometry for deoxyribonucleic acid and intermediate filament proteins of residual mature teratoma: all tumor cells are aneuploid. Lab Invest 64:113–117

Looijenga LHJ, Gillis AJM, Van Putten WLJ, Oosterhuis JW (1993) In situ numeric analysis of centromeric regions of chromosomes 1, 12, and 15 of seminoma, nonseminomatous germ cell tumors, and carcinoma in situ of human testis. Lab Invest 68:211–219

Manivel CJ, Jessurum J, Wick MR, Dehar LP (1987) Placental immunoreactivity in testicular germ cell neoplasms. Am J Surg Pathol 11:21–29

Oosterhuis JW, Castedo SMMJ, De Jong B, Cornelisse CJ, Dam A, Sleijfer DTH, Schraffordt Koops H (1989) Ploidy of subtypes of primary gern cell tumors of the testis: pathogenetic and clinical relevance. Lab Invest 60:14–21

Skakkebaek NE, Berthelsen JG, Giwercman A, Müller J (1987) Carcinoma-in-situ of the testis (CIS) possible origin from gonocytes, and precursor of all types of germ cell tumours except spermatocytoma. Int J Androl 10:19–28

Vos A, Oosterhuis JW, De Jong B, Buist J, Schraffordt Koops H (1990) Cytogenetics of carcinoma in situ of the testis. Cancer Genet Cytogenet 46:75–81

Walt H, Emmerich P, Cremer T, Hofmann MC, Bannwart (1989) Supernumerary chromosome 1 in interphase nuclei of atypical germ cells in paraffin-embedded human seminiferous tubules. Lab Invest 61:527–531

Cytogenetic and Molecular Studies of Clear-Cell Sarcoma of Tendons and Aponeuroses

G. Stenman[1], P. Sahlin[1], L. Angervall[2], P. Meltzer[3], and L.-G. Kindblom[2]

1 Introduction

Clear-cell sarcoma of tendons and aponeuroses (CCS) was first described and named by Enzinger (1965) and is currently recognized as a rare and distinct clinicopathologic entity among soft tissue sarcomas. Originally it was suspected that there was a relationship between CCS and synovial sarcoma (Enzinger 1965; Kubo 1969; Hajdu et al. 1977). The light-microscopic similarity to malignant melanoma, the occurrence of melanin in some cases (Enzinger 1965), and the ultrastructural finding of melanosomes (Hoffman and Carter 1973) led Chung and Enzinger (1983) to suggest that the term clear-cell sarcoma of tendons and aponeuroses should be changed to malignant melanoma of soft parts. However, melanocytic features are not consistently found, and differentiation along Schwann's cell lines has also been observed (Kindblom et al. 1983; Azumi and Turner 1983; Ohno et al. 1986). The present knowledge of CCS, including electron microscopy, immunocytochemistry and cytochemistry, indicates that this entity is of neural crest origin presenting variable degrees of differentiation along melanocytic and Schwann's cell lines.

2 Clinicopathologic Characteristics of CCS

CCS shows a predilection for the extremities, especially the region of the foot, ankle, knee, thigh, and hand. The tumor is usually deep seated, and, as the name indicates, it is usually intimately associated with tendons and aponeuroses. Clinically, the tumor presents as a slowly growing mass, causing pain in less than half of the cases. A very long duration of symptoms

[1] Laboratory of Cancer Genetics, Department of Pathology, Sahlgrenska University Hospital, S-41345 Gothenburg, Sweden
[2] Department of Pathology, Sahlgrenska University Hospital, Gothenburg, Sweden
[3] Department of Pediatrics and Radiation Oncology, Division of Cancer Biology, University of Michigan Medical Center, Ann Arbor, Michigan, USA

has been noted in many cases, sometimes up to 20 years (Enzinger 1965). Grossly, the tumor appears well circumscribed, although it is not encapsulated, and the size ranges from 1 cm to more than 10 cm, with the majority of the tumors measuring less than 6 cm in diameter.

Microscopically, the tumor is characterized by nests and fascicles of rounded or fusiform cells, often having a clear cytoplasm and characteristically a vesicular nucleus with a very prominent nucleolus. The cell nests and fascicles are divided by a prominent framework of fibrocollagenous tissue, which appears to be continuous with adjacent tendons or aponeuroses. Scattered multinucleated giant cells with peripherally placed nuclei are frequently found. Melanin can rarely be demonstrated in routinely stained sections but with sensitive melanin stainings such as Whartin-Starry silver staining at low pH (3.2) up to 75% of the cases have been reported to contain intracytoplasmic melanin (Enzinger and Weiss 1988). CCS reveals immunoreactivity for a number of neural markers such as neuron specific enolase, S-100 protein, Leu-7, and synaptophysin, as well as melanoma-associated antigens revealed by the antibodies HMB-45 and NKI/C3. Cytochemical studies have also shown expression of melanin-related metabolic pathways involving thyrosinase and cholinesterase activities in CCS (Mii et al. 1989). Ultrastructurally the tumor cells of CCS have been found to show not only melanocytic features, including melanosomes (Hoffman and Carter 1973; Kindblom et al. 1983), but also Schwann's cell features with the formation of external laminae, interdigitating cytoplasmic projections and the occurrence of so-called long-spaced collagen (Azumi and Turner 1983; Kindblom et al. 1983; Ohno et al. 1986).

The long-term prognosis for CCS is poor, and the majority of patients eventually die of metastatic disease. The course is, however, often protracted and local recurrences may appear late. The most common sites of metastases are the lung, regional lymph nodes, and bone (Chung and Enzinger 1983; Chung 1991).

3 Case Report

A 36-year-old man sought medical care for a painless tumor at the lateral aspect of the right knee joint, noticed for several years but observed to be slowly increasing in size over the past 6 months (Stenman et al. 1992). The 8 × 8 cm large tumor was removed by a wide local excision. Grossly, the tumor was rather firm and revealed gray-white, uniformly cut surfaces. Light- and electron-microscopically the tumor had an appearance characteristic of CCS (Figs. 1a,b, 2a,b). Immunocytochemically the tumor cells revealed a strong reactivity for vimentin, S-100 protein (Fig. 1c) and the melanoma-associated antigens detected by the antibodies NKI/C3 and HMB-45. There was also immunoreactivity for laminin and collagen IV,

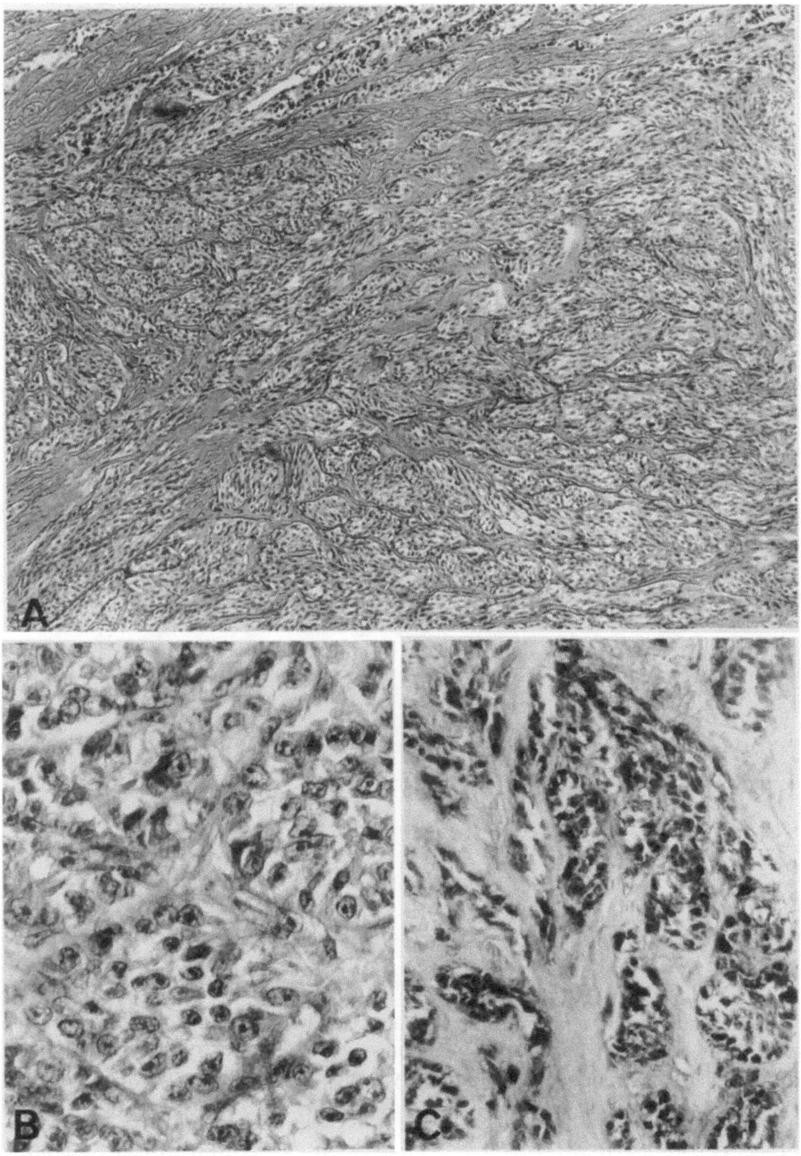

Fig. 1. **A** Clear-cell sarcoma characterized by tumor cells arranged in nests, which split up the aponeurotic tissue of a tendon. **B** The tumor cells show a poorly outlined pale cytoplasm and oval or round nuclei with prominent nucleoli. **C** The tumor cells show a strong immunoreactivity for S-100 protein (avidin-biotin complex technique)

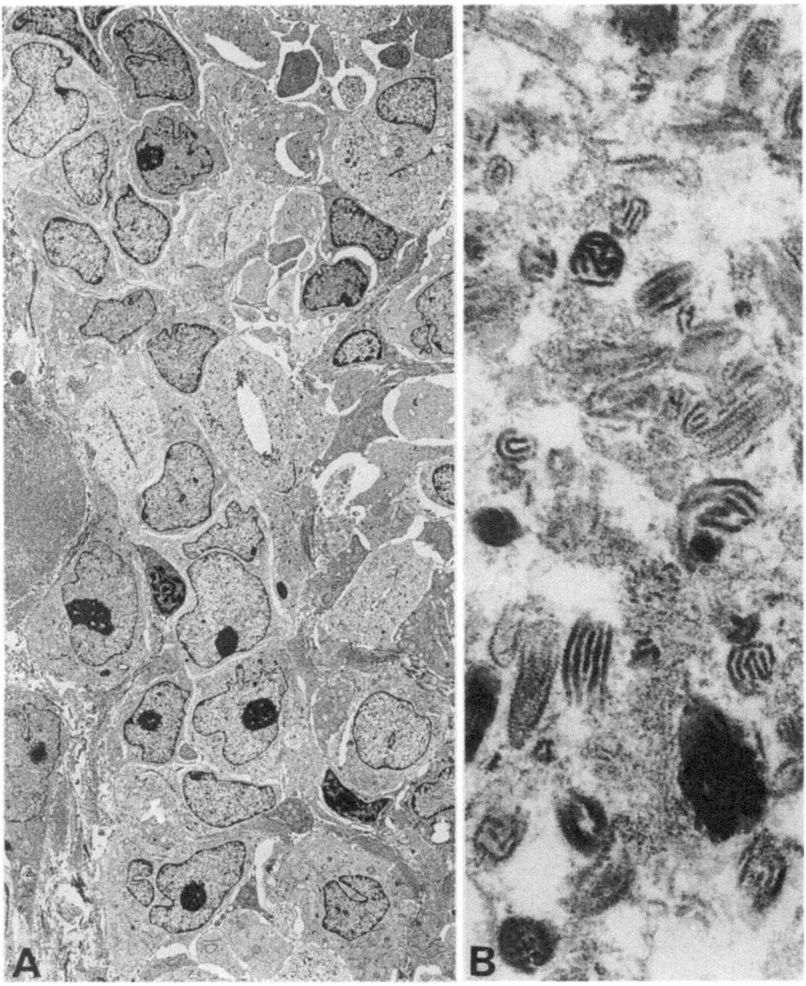

Fig. 2. **A** Ultrastructurally, the tumor cells are characterized by irregularly shaped nuclei with evenly dispersed chromatin and a single prominent nucleoli (×1750). **B** Cytoplasmic detail of a tumor cell with numerous melanosomes showing varying degrees of differentiation (×40 000)

indicating that external lamina material surrounds individual tumor cells and cell groups.

4 Cytogenetic and In Vitro Growth Characteristics of CCS

Primary cultures were established from collagenase treated fresh tumor tissue as previously described (Stenman et al. 1992). Cells were harvested for cytogenetic and morphologic analyses after 12 days in vitro.

The monolayer of cultured tumor cells was characterized by stellate and dendritic cells having a single oval or rounded nucleus with one or more prominent nucleoli. There were also some scattered, irregularly shaped, very large, cytoplasm-rich cells. The tumor cells tended to adhere in small clusters or were arranged in anastomosing rows. On ultrastructural examination, very few oval or rod-shaped membrane bound bodies were identified, which could represent premelanosomes. The cultured tumor cells revealed a strong immunoreactivity for vimentin and S-100 protein. Some 25% of the cells were positively stained using the HMB-45 melanoma antibody.

Cytogenetic analysis revealed a hyperdiploid stemline with the karyotype 49,XY,der(1)t(1;?)(p36.1–36.3;?),+8,+8,+12,t(12;22)(q13;q13) (Stenman et al. 1992). The t(12;22) was the only abnormality present in all cells (Fig. 3), whereas the 1p+ marker was detected in about 76% of the cells. Extra copies of chromosomes 8 and 12 were found in 94% and 88% of the cells, respectively. There were also two sidelines, one of which differed from the stemline by a gain of one or two chromosomes 7, and the other by a duplication of parts of 7q, dup(7)(q11.2 → q36).

Only few CCS have been cytogenetically characterized. Including the present case, data is available from a total of eight cases (Bridge et al. 1990, 1991; Peulvé et al. 1991; Stenman et al. 1992; Speleman et al. 1992; Rodriguez et al. 1992). The cytogenetic findings in these cases are summarized in Table 1. Cases 1, 2, 3, 5, and 6 represent primary lesions, while cases 4 and 7 are recurrencies and case 8 a metastatic lesion. All but two of the tumors had near-diploid stemlines. The karyotypes of most tumors were very complex. The most consistent abnormality observed among the eight cases was an apparently identical t(12;22)(q13;q13) found in six tumors. In

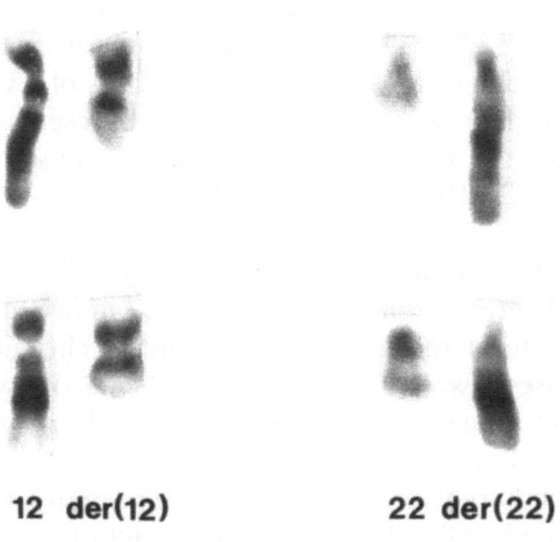

Fig. 3. Partial karyotypes showing the t(12;22)(q13;q13) typical of CCS

Table 1. Summary of clonal abnormalities in eight cases of CCS

Case no.	Karyotype(s)	Reference
1	46,XY,+8,**t(12;22)(p11.2;p11.2)**,t(15;15)(p13;q21),−22	Bridge et al. (1990)
2	66,XY,+X,+i(1q),+2,+3,+del(6)(q12),+der(7)t(7;?)(q35;?), +der(7)t(7;?)(q35;?),+8,+8,+13,+13,+14,+15,+15,+del(16) (q12),+17,+der(19)t(19;?)(q13;?),+der(20)t(20;?)(q13;?),+21, +22	Bridge et al. (1990)
3	50,XX,t(2;?)(q35;?),+7,+8,+8,+18,t(22;?)(q11;?)	Bridge et al. (1991)
4	39−42,XY,−2,del(4)(q26.1),−5,+der(5)t(5;15)(p15.3;p13),−6,−7, −9,der(9)t(2;9)(q11.2;p22),**t(12;22)(q13;q13)**,−14,−15,−16,−17, +1−3mar [cp5]	Bridge et al. (1991)
5	46,XX,**t(12;22)(q13−14;q12)**/44,XX,−1,+dic(1;?;9)(1qter→ 1p21−22::?::9p22−24→9qter),−9,−9,−12,−22,−22,**+der(22)t (12;22)(q13−14;q12)**,+2mar,+min	Peulvé et al. (1991)
6	49,XY,der(1)t(1;?)(p36.1−36.3;?),+8,+8,+12,**t(12;22)(q13; q13)**	Stenman et al. (1992)
7	48−51,XY,del(1)(q42),+i(1q),+i(1q),−3,+7,+8,**t(12;22)(q13; q12−13)**,+17,−18,−21 [cp5]	Speleman et al. (1992)
8	65−101,XXY,+X,−Y,del(2)(p11),del(5)(q22),−6,i(6)(p10),+8,+8, +9,−11,−11,**+t(12;22)(q13;q13)**,−14,−15,−17,−17,+3mar [cp4]	Rodriguez et al. (1992)

one case, the t(12;22) was the sole abnormality (Peulvé et al. 1991). In the present case, detailed analysis of the karyotypic evolution indicated that the t(12;22) was the first step in the evolution of the abnormal stemline. Taken together, these findings strongly indicate that the t(12;22)(q13;q13) is a primary cytogenetic abnormality in CCS. To the best of our knowledge this translocation has not been found in any other neoplastic disorder (Mitelman 1991) and can therefore be regarded as diagnostic for this tumor entity. Interestingly, case 3 had a t(22;?)(q11;?) where the material translocated onto 22q was unknown (Bridge et al. 1991). As more cases of CCS are being studied it will become clear whether this in fact represents a variant translocation. Case 2 was the only tumor without structural rearrangements involving 12q and/or 22q (Bridge et al. 1990).

Nonrandom, secondary changes of possible progressional importance were found in all seven cases of CCS. Additional copies of normal chromosomes 8 and normal or rearranged chromosomes 7 were observed in five and four cases, respectively, and structural rearrangements involving chromosome 1 were detected in four cases, two of which contained one or two i(1q). Abnormalities involving all three chromosomes were seen in two cases. Most of the cases also contained random numerical and structural changes. The following steps in the karyotypic evolution of the present CCS were either observed or could be deduced: t(12;22)(q13;q13), gain of one or two chromosomes 8, gain of one chromosome 12, t(1;?)(p36.1−36.3;?), gain of one or two chromosomes 7 and/or dup(7)(q11.2→q36). Follow-up studies of cases with these nonrandom abnormalities will indicate their role in the prognosis of CCS.

The histogenesis of CCS has been a matter of debate (see Sect. 1). However, current knowledge does indicate that CCS is of neural crest derivation, presenting variable degrees of differentiation along melanocytic and Schwann's cell lines. Interestingly, other tumors of possible neural crest origin, Ewing's sarcoma, Askin's tumor, and neuroepithelioma, also show a consistent translocation involving 22q, i.e., a t(11;22)(q24;q12) (Aurias et al. 1983; Turc-Carel et al. 1988a). Whether in fact the t(12;22) in CCS and the t(11;22) in Ewing's sarcoma involve the same locus on 22q is at present unknown. There are also reports of other sarcomas with translocations of 22q, including extraskeletal myxoid chondrosarcoma with t(9;22)(q31;q12.2) (Turc-Carel et al. 1988b) and hemangiopericytoma with t(13;22)(q22;q11) (Limon et al. 1986). The chromosomal pattern in CCS also bears some resemblance to that in malignant melanoma. As pointed out by Bridge et al. (1990, 1991), abnormalities involving chromosomes 1, 2, 6 and 7, frequently seen in malignant melanoma, may also be found in CCS (Table 1). However, as suggested for malignant melanoma, these deviations might reflect progression towards more advanced stages of the disease, rather than being primary abnormalities of pathogenetic importance. Since the t(12;22)(q13;q13) has never been encountered in malignant melanoma or in any other neoplastic condition, this rearrangement must be considered specific for CCS, and thus distinguishes this tumor type from malignant melanoma.

The 12q13–15 region appears to be a hot spot for chromosomal rearrangements in tumors. In addition to CCS, the following tumors have shown consistent rearrangements with breakpoints in this region: pleomorphic adenomas of the salivary gland (Bullerdiek et al. 1987; Sandros et al. 1990), lipomas (Sreekantaiah et al. 1991), myxoid liposarcomas (Turc-Carel et al. 1986), and uterine leiomyomas (Nilbert and Heim 1990). Whether there is a common breakpoint on 12q in these tumors is uncertain. From a cytogenetic point of view it appears as if the breakpoints might differ both within the different tumor types and among the different diagnostic entities. For a definite answer to this question we must await molecular cloning of the various 12q rearrangements.

5 Molecular Studies of the 12q13 Translocation Breakpoint

The breakpoints in the t(12;22)(q13;q13) in CCS are likely to harbor genes important in the regulation of cellular proliferation and/or differentiation. Since the der(12) chromosome have been found lost in several cases these genes are most likely located on the der(22) chromosome proximal to the 22q13 breakpoint and distal to the 12q13 breakpoint (Stenman et al. 1992). Several candidate genes have been mapped to the 12q13–14 region, in-

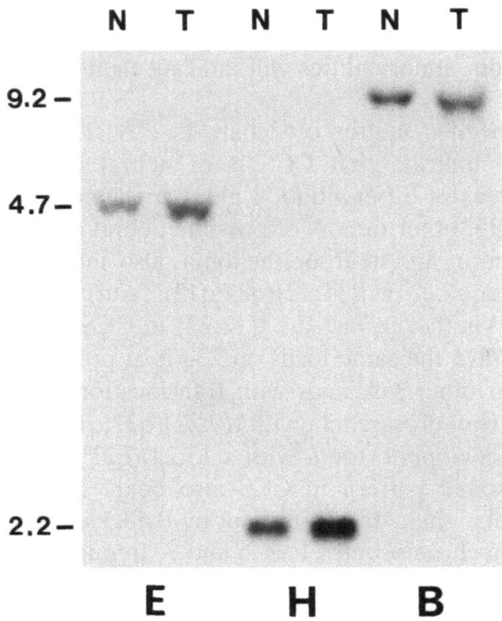

Fig. 4. Southern blot analysis of *SAS* in CCS. Normal (*N*) and tumor (*T*) DNAs were digested with *Eco*RI (*E*) *Hind*III (*H*), or *Bam*HI (*B*), blotted and hybridized to a SAS probe (Meltzer et al. 1991). *Left*, the fragment sizes (in kb)

cluding *GLI* (12q13), *WNT1* (12q13), *COL2A1* (12q13), *ERBB3* (12q13), *SAS* (12q13–14) and *A2MR* (12q13–14) (Craig and McBride 1991). By southern blot analysis we have tested these six genes for rearrangements in the present case of CCS. No evidence of rearrangement, deletion, or amplification of any of the genes was found in tumor DNAs digested with *Eco*RI, *Hind*III, or *Bam*HI (Fig. 4), thus indicating that none of the genes seem to be the primary target gene on 12q. However, the analyses do not rule out the possibility of rearrangements located outside the DNA sequences analyzed. These findings are in accordance with recent molecular studies of several other tumors with 12q changes (revieved in Sahlin et al. 1992 and Åman et al. 1992), including lipomas, myxoid liposarcomas, leiomyomas, and pleomorphic adenomas, where one or more of these genes have been tested, and no aberrant restriction pattern have been found.

6 Concluding Remarks

In this chapter we review the chromosomal findings in eight cases of CCS. The most consistent abnormality observed was an apparently identical t(12;22)(q13;q13) found in six tumors. It is concluded that this translocation represents a primary, tumor-specific abnormality diagnostic for this entity. Due to light-microscopic and ultrastructural similarities to malignant melanoma it has previously been suggested that the term CCS should be changed to malignant melanoma of soft parts. However, since CCS shows ultrastrutural evidence of differentiation also along Schwann's cell lines, and the t(12;22)(q13;q13) has never been encountered in malignant melanoma, we suggest that the term CCS should be retained as a descriptive label for this neuroectodermal sarcoma. Our molecular studies of the 12q13 breakpoint in CCS allowed us to exclude the following candidate genes as the target for the t(12;22): *GLI*, *WNT1*, *COL2A1*, *ERBB3*, *SAS*, and *A2MR*. It is likely that novel, possibly tissue and/or cell specific, genes are to be found at the translocation breakpoints.

Acknowledgements. We thank Matthias Kraus, Ola Myklebost, Bert Vogelstein and Eero Vuorio for probes. This study was supported by grants from the Inga Britt and Arne Lundbergs Research Foundation, Assar Gabrielssons Cancer Research Foundation, the Jubileumsklinikens Cancer Research Foundation, the Gothenburg Medical Society, and the Swedish Cancer Society.

Note Added in Proofs. After submission of this manuscript Zucman et al. (Nature Genet 4:341–345, 1993) reported the cloning of the genes involved in the t(12;22) in CCS. The translocation was shown to result in the fusion of the transcription factor gene ATF1 on chromosome 12 to the EWS (Ewing Sarcoma) gene on chromosome 22. The EWS gene encodes a protein with a putative RNA-binding domain in the C-terminal portion which in the fusion protein is replaced by the DNA-binding domain of ATFI.

References

Åman P, Ron D, Mandahl N, Fioretos T, Heim S, Arheden K, Willén H, Rydholm A, Mitelman F. Rearrangement of the transcription factor gene CHOP in myxoid liposarcoma with t(12;16)(q13;p11). Genes Chrom Cancer 5:278–285

Aurias A, Rimbaut C, Buffe C, Dubousset J, Mazabraud A (1983) Chromosomal translocations in Ewing's sarcoma. N Engl J Med 309:496–497

Azumi N, Turner RR (1983) Clear cell sarcoma of tendons and aponeuroses: electron microscopic findings suggesting Schwann cell differentiation. Hum Pathol 14:1084–1089

Bridge JA, Borek DA, Neff JR, Huntrakoon M (1990) Chromosomal abnormalities in clear cell sarcoma. Implications for histogenesis. Am J Clin Pathol 93:26–31

Bridge JA, Screekantaiah C, Neff JR, Sandberg AA (1991) Cytogenetic findings in clear cell sarcoma of tendons and aponeuroses. Malignant melanoma of soft parts. Cancer Genet Cytogenet 52:101–106

Bullerdiek J, Bartnitzke S, Weinberg M, Chilla R, Haubrich J, Schloot W (1987) Rearrangements of chromosome region 12q13–15 in pleomorphic adenomas of the human salivary gland (PSA). Cytogenet Cell Genet 45:187–190

Chung EB (1991) Clear cell sarcoma: a tumor of probable neuroectodermal origin. In: Ninfo V, Chung EB, Cavazzana AO (eds) Tumors and tumorlike lesions of soft tissue. Churchill Livingstone, New York, pp 259–263

Chung EB, Enziner FM (1983) Malignant melanoma of soft parts; a reassessment of clear cell sarcoma. Am J Surg Pathol 39:405–413

Craig IW, McBride OW (1991) Report of the committee on the genetic constitution of chromosome 12. Human gene mapping 11 (1991). 11th international workshop on human gene mapping. Cytogenet Cell Genet 58:555–579
Enzinger FM (1965) Clear-cell sarcoma of tendons and aponeuroses: an analysis of 21 cases. Cancer 18:1163–1174
Enzinger FM, Weiss SW (1988) Soft tissue tumors. Mosby, St. Louis, pp 945–951
Hajdu SI, Shiu MH, Fortner JG (1977) Tendosynovial sarcoma: a clinicopathological study of 136 cases. Cancer 42:1201–1217
Hoffman GJ, Carter D (1973) Clear-cell sarcoma of tendons and aponeuroses with melanin. Arch Pathol 95:22–25
Kindblom L-G, Lodding P, Angervall A (1983) Clear-cell sarcoma of tendons and aponeuroses. An immunohistochemical and electron microscopic analysis indicating neural crest origin. Virchows Arch (Pathol Anat) 409:109–128
Kubo T (1969) Clear-cell sarcoma of patellar tendon studied by electron microscopy. Cancer 24:948–953
Limon J, Rao U, Dal Cin P, Gibas Z, Sandberg AA (1986). Translocation (13;22) in a hemangiopericytoma. Cancer Genet Cytogenet 21:309–318
Meltzer PS, Jankowski SA, Dal Cin P, Sandberg AA, Paz IB, Coccia MA (1991) Identification and cloning of a novel amplified DNA sequence in a human malignant fibrous histiocytoma derived from a region of chromosome 12 frequently rearranged in soft tissue tumors. Cell Growth Differ 2:1–7
Mii Y, Miyauchi Y, Hohnoki K, Maruyama H, Tsutsumi M, Dohmae K, Tamai S, Konishi Y, Yamanouchi T (1989) Neural crest origin of clear cell sarcoma of tendons and aponeuroses: ultrastructural and enzyme cytochemical study of human and nude mouse-transplanted tumors. Virchows Arch [A] 415:51–60
Mitelman (1991) Catalog of chromosome aberrations in cancer. Wiley-Liss, New York
Nilbert M, Heim S (1990) Uterine leiomyoma cytogenetics. Genes Chrom Cancer 2:3–13
Ohno T, Park P, Utsunomiya Y, Hirahata H, Inoue K (1986) Ultrastructural study of a clear cell sarcoma suggesting schwannoma differentiation. Ultrastruct Pathol 10:39–48
Peulvé P, Michot C, Vannier J-P, Tron P, Hemet J (1991) Clear cell sarcoma with t(12;22)-(q13–14;q12). Genes Chrom Cancer 3:400–402
Rodriguez E, Sreekantaiah C, Reuter VE, Motzer RJ, Chaganti RSK (1992) t(12;22)(q13;q13) and trisomy 8 are nonrandom aberrations in clear-cell sarcoma. Cancer Genet Cytogenet 64:107–110
Sahlin P, Mark J, Stenman G (1992) The INT1 and GLI genes are not rearranged or amplified in benign pleomorphic adenomas with chromosome abnormalities of 12q13–15. Cancer Genet Cytogenet 58:85–88
Sandros J, Stenman G, Mark J (1990) Cytogenetic and molecular observations in human and experimental salivary gland tumors. Cancer Genet Cytogenet 44:153–167
Speleman F, Colpaert C, Goovaerts G, Leroy JG, Van Marck E (1992) Malignant melanoma of soft parts – further cytogenetic characterization. Cancer Genet Cytogenet 60:176–179
Sreekantaiah C, Leong SPL, Karakousis CP, McGee DL, Rappaport WD, Villar HV, Neal D, Fleming S, Wankel A, Herrington PN, Carmona R, Sandberg AA (1991) Cytogenetic profile of 109 lipomas. Cancer Res 51:422–433
Stenman G, Kindblom L-G, Angervall A (1992) Reciprocal translocation t(12;22)(q13;q13) in clear-cell sarcoma of tendons and aponeuroses. Genes Chrom Cancer 4:122–127
Turc-Carel C, Limon J, Dal Cin P, Rao U, Karakousis C, Sandberg AA (1986) Cytogenetic studies of adipose tissue tumors. II. Recurrent reciprocal translocation t(12;16)(q13;p11). Cancer Genet Cytogenet 23:291–299
Turc-Carel C, Aurias A, Mugneret F, Lizard Sidaner I, Volk C, Thiery JP, Olschwang S, Philip I, Berger MP, Philip T, Lenoir GM, Mazabraud A (1988a) Chromosomes in Ewing's sarcoma. An evaluation of 85 cases and remarkable consistency of t(11;22)(q24;q12). Cancer Genet Cytogenet 32:229–238
Turc-Carel C, Dal Cin P, Rao U, Karakousis C, Sandberg AA (1988b) Recurrent breakpoints at 9q31 and 22q12.2 in extraskeletal myxoid chondrosarcoma. Cancer Genet Cytogenet 30:145–150

The Molecular Genetics of Chromosome 12

Comparative Gene Mapping: Human Chromosome 12 and Mouse Chromosome 15

H. Hameister, C. Klett, G. Hartmann, and C. Ebensperger

In the past we have mapped 11 genes and DNA sequences on mouse chromosome 15 by in situ hybridization and by linkage analysis (Adolph et al. 1987, 1988, 1989). The combined data of these studies are shown in Fig. 1. Most of mouse chromosome 15 is conserved in three linkage groups in the human genome. These linkage groups reside on human chromosomes 8q24, 22q13.3-qter and 12cen-q15 (Fig. 1). The linkage group common to mouse chromosome 15 and human chromosome 12 comprises the loci homeo box region 3 (HOX3); wingless-type MMTV integration site 1 (WNT1); and glycerol-3-phosphate dehydrogenase (GPD1). Of these genes GPD1 attracted our interest. This is a biochemical marker for adipocytes (Wise and Green 1979). The enzyme provides fat tissue with free glycerol for triglyceride synthesis. The enzyme is highly expressed in mature adipocytes. Benign lipomas and liposarcomas are characterized by chromosomal translocations. One breakpoint which is most common to these tumors is localized in the region 12q13–15 (see Sreekantaiah et al. 1991, for overview). It was therefore of interest to map the gene GPD1 as exactly as possible in the human karyotype. Figure 2 shows the results of an in situ hybridization analysis. GPD1 is localized on chromosome 12 in the region 12q13.

During our continuous efforts to contribute to the mapping of the human genome the following further genes were mapped by in situ hybridization to human chromosome 12: mannose-6-phosphate receptor (cation dependent; M6PR), 12p12–13; GPD1, 12q13; myogenic factor 5 (MYF5), 12q21; myogenic factor 6 (MYF6), 12q21; proteoglycan (decorin; DCN), 12q21.3. These date are summarized in Fig. 3. M6PR recognizes mannose-6-phosphate residues which are specifically added to newly synthesized lysosomal enzymes and are transported by the mannose-6-phosphate receptor from the Golgi apparatus to the lysosomes (Pohlmann et al. 1987). The gene has been mapped to human chromosome 12 by soma cell hybrid analysis in a foregoing study (Pohlmann et al. 1987). DCN belongs to the class of proteoglycans which interconnect the collagen fibrils in the extracellular matrix. The localization as reported here has been published (Just et

Abteilung Klinische Genetik, Universität Ulm, 89073 Ulm, Germany

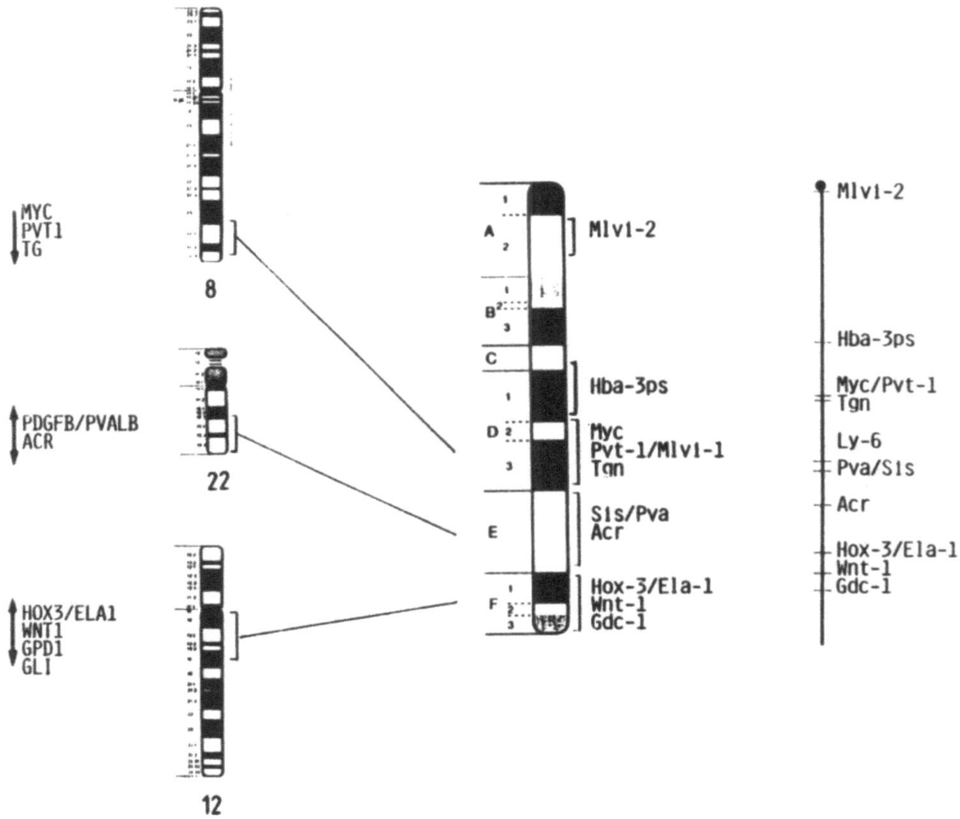

Fig. 1. Mapping data from mouse chromosome 15 and the linkage groups conserved in the human genome. *Middle*, ideogram of MMU 15; the genes localized by in situ hybridization are indicated. *Right*, linear order of these loci as determined by linkage analysis. *Left*, areas of linkage conservation within human chromosomes and the respective gene loci. *Arrow*, the orientation of these linkage groups in human

al. 1991). Special interest attaches to the localization of the two myogenic factors MYF5 and MYF6 in 12q21. Both of these genes were isolated due to their sequence similarity to the murine myogenic determination factor MyoD1 (Braun et al. 1990). These factors are expressed in skeletal muscle tissue and are able to convert competent fibroblasts to the muscle phenotype. MYF5 and MYF6 are localized in the same transcriptional orientation nearly 6.5 kb next to each other (Braun et al. 1990). Similar to the involvement of the gene GPD1 in a breakpoint analysis of lipomas, it might be of interest to investigate MYF5 and MYF6 rearrangements in uterine leiomyomas which, similar to lipomas, most often bear a breakpoint in the region 12q13–15 (Pandis et al. 1990). Due to inherent cytogenetic inprecision the break and the localization as found by our analysis may overlap.

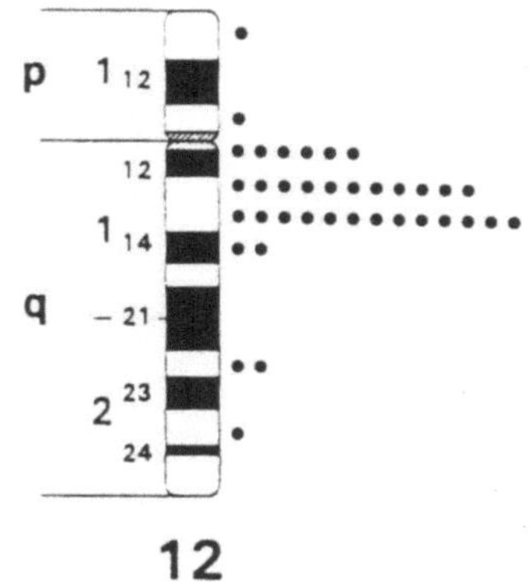

Fig. 2. Results of in situ hybridization with a genomic 4.0 kb *Hin*dIII fragment from the GPD1 gene (Gwynn et al. 1990). Grain accumulation is shown for 12q13

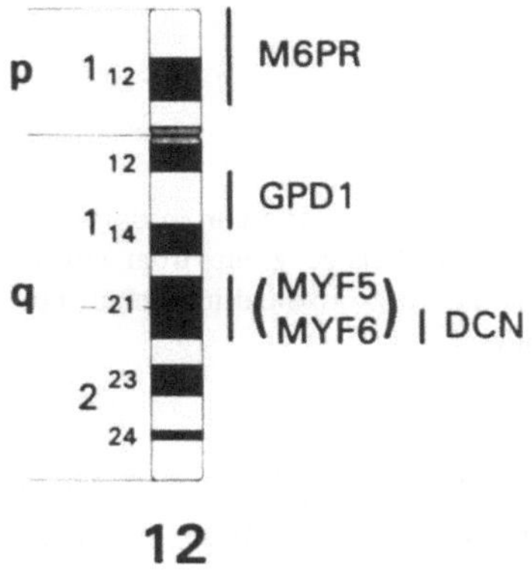

Fig. 3. Summary of our in situ hybridization results on human chromosome 12 with different gene probes. The probe for M6PR was provided by Dr. R. Pohlmann (Göttingen), that for MYF5 and MYF6 by Dr. H.H. Arnold (Hamburg), and that for (DCN) by Dr. U. Vetter (Bethesda)

The genes mapped during this study on human chromosome 12 are dispersed over the whole of that chromosome. They belong, again, to three linkage groups conserved in the mouse genome (Searle et al. 1989; Davisson et al. 1991). These linkage groups are found on MMU 6, MMU 10, and MMU 15 (Fig. 4). The mouse gene map is based mainly on linkage data which provides the investigator with genetic data and information about the linear order of the respective loci along the chromosome. The human gene map is based mostly on physical mapping data which are provided by soma cell hybrid analysis and in situ hybridization. These methods allow the assignment of the respective loci directly to the chromosomal banding pattern. However, due to the poor resolution power of cytogenetic techniques information on the exact order is scarce up to now for individual chromosomal subsections in the human genome. So far mouse and human gene mapping data complement each other. An example for this is given in Fig. 4.

The linkage group common to human chromosome 12 and mouse chromosome 6 is built up by 8 loci and extends to 18 cM in the mouse. Ly-4 (human CD4) is orientated most proximal and Kras-2 most distal. In human

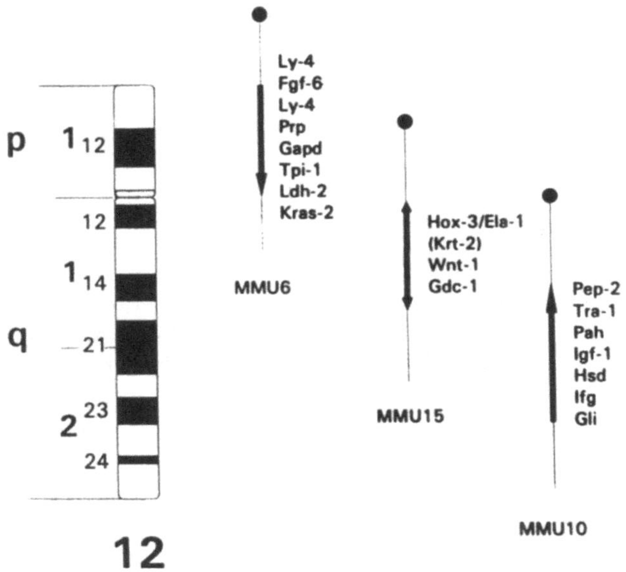

Fig. 4. Human chromosome 12 and the linkage groups conserved on mouse chromosomes 6, 10, and 15. *Arrow*, the orientation of the linkage groups. The linkage group from mouse chromosome 10 must be turned upside down, with GLI1 as the most proximal in human. For HOX3-GPD1 the orientation in human is doubtful up to now

KRAS2 is localized most proximal in 12p12.1, LDHB in 12p12.1–p12.2 and all other genes in 12p13. This allows one to orientate this linkage group in human with KRAS2 most proximal to the centromere. The order of loci as determined in mouse can now be transferred to the human gene map. This is most effective for the distal six genes localized in human chromosome 12p13 (Fig. 4). Another informative example provides the linkage group common to human chromosome 12 and mouse chromosome 10. This linkage group comprises at least seven loci which in the mouse extend over 27 cM and are ordered from cen . . . Pep-2, Tra-1, Pah, Igf-1, Hsd, Ifg, Gli . . . ter. In human GLI1 marks the most proximal locus in the region 12q13 and TRA1 the most distal locus in 12q24.2-qter (Maki et al. 1991). The comparative mapping data raise considerable doubt as to the localization of PEPB to 12q21. The orientation is upside down in human (indicated by an arrow in Fig. 4). Again, the exact order known in mouse can be transferred to the human gene map. The human 12qter region is conserved on mouse chromosome 5. Only two loci common to both, HSA 12 and MMU 5, are known (ACADS and TCF1).

The orientation of the third conserved linkage group on mouse chromosome 15 has been determined by our studies as shown in Fig. 1. Other studies make the exact orientation questionable (Watanabe et al. 1990; Saffer et al. 1990). This linkage group extends in the mouse to more than 20 cM. In human information is not available to arrange these loci in a center orientation. In humans GLI1 has also been localized to 12q13. As GLI1 belongs to the linkage group found on MMU10, it is evident that GLI1 in human maps distal from HOX3, WNT1, and GPD1.

Several mutations in mouse have shown to be the exact correlate of known human diseases (Tassabehji et al. 1992; Fleischman et al. 1991). The region of human chromosome 12q13–15 bears a gene locus engaged in multiple forms of human mesenchymal tumors. Therefore it seems interesting to consider whether one of the mutations mapped in the regions on MMU 15 and MMU 10 which are conserved on human chromosome 12 show mesenchymal or mesodermal growth disturbance. A candidate gene is the long-known and well-described mutation steel (Sl). It has been shown that Sl encodes a growth factor which acts as a ligand for the c-*kit* oncogene receptor (Flanagan et al. 1991).

Acknowledgement. This research was supported by a grant from the Deutsche Forschungsgemeinschaft.

References

Adolph S, Bartram CR, Hameister H (1987) Mapping of the oncogenes Myc, Sis and int-1 to the distal part of mouse chromosome 15. Cytogenet Cell Genet 44:65–68

Adolph S, Strauss PG, Hameister H (1988) Localization of proviral integration sites (Mlvi-1, Mlvi-2, and Pvt-1) and the α-globin pseudogene, Hba-3ps, on murine chromosome 15. Cytogenet Cell Genet 47:189–191

Adolph S, Berchtold MW, Hameister H (1989) Fine localization of genes on distal murine chromosome 15. Cytogenet Cell Genet 52:177–179

Braun T, Bober E, Winter B, Rosenthal N, Arnold H (1990) Myf-6, a new member of the human gene family of myogenic determination factors: evidence for a gene cluster on chromosome 12. EMBO J 9:821–831

Davisson MT, Lalley PA, Peters J, Doolittle DP, Hillyard AL, Searle AG (1991) Report of the comparative committee for human, mouse and other rodents. Human Gene Mapping 11. Cytogenet Cell Genet 58:1152–1189

Flanagan JG, Chan DC, Leder P (1991) Transmembrane form of the kit ligand growth factor is determined by alternative splicing and is missing in the Sl^d mutant. Cell 64:1025–1035

Fleischman RA, Saltman DL, Stastny V, Zneimer S (1991) Deletion of the c-kit protooncogene in the human developmental defect piebald trait. Proc Natl Acad Sci USA 88:10885–10889

Gwynn B, Liford KA, Birkenmeier EH (1990) Sequence conservation and structural organization of the glycerol-3-phosphate dehydrogenase promoter in mice and humans. Mol Cell Biol 10:5244–5256

Just W, Vetter U, Fischer LW, Klett C, Vogel W (1991) In situ localization of decorin and biglycan on chromosomes 12 and X. Human Gene Mapping 11. Cytogenet Cell Genet 58:1977–1978

Maki RG, Eddy RL, Byers MG, Shows TB, Srivastava PK (1991) Localization of genes for the stress-inducible tumor rejection antigen gp96. Human Gene Mapping 11. Cytogenet Cell Genet 58:1978–1979

Pandis N, Heim S, Bardi G, Mandahl N, Mitelman F (1990) High resolution mapping of consistent leiomyoma breakpoints in chromosomes 12 and 14 to 12q15 and 14q24.1. Genes Chrom Cancer 2:227–230

Pohlmann R, Nagel G, Schmidt B, Stein M, Lorkowski G, Krentler C, Cully J, Meyer HE, Grezeschik KH, Meersmann G, Hasilik A, Von Figura K (1987) Cloning of a cDNA encoding the human cation-dependent mannose 6-phosphate-specific receptor. Proc Natl Acad Sci USA 84:5575–5579

Saffer JD, Thurston SJ, Annarella MB, Compton JG (1990) Localization of the gene for the trans-acting transcription factor Spl to the distal end of mouse chromosome 15. Genomics 8:571–574

Searle AG, Peters J, Lyon MF, Hall JG, Evans EP, Edwards JH, Buckles VJ (1989) Chromosome maps of man and mouse. IV. Ann Hum Genet 53:89–140

Sreekantaiah C, Leong SPL, Karakousis CP, McGee DL, Rappaport WD, Villar HV, Neal D, Fleming S, Wankel A, Herrington PN, Carmona R, Sandberg AA (1991) Cytogenetic profile of 109 lipomas. Cancer Res 51:422–433

Tassabehji M, Read PA, Newton VE, Harris R, Balling R, Gruss P, Strachan T (1992) Waardenburg's syndrome patients have mutations in the human homologue of the Pax-3 paired box gene. Nature 355:635–636

Watanabe T, Ohno K, Shimizu A, Sakai Y, Takahashi M, Takahashi N (1990) Mapping of the Hox-3.1 and Myc-1.2 genes on chromosome 15 of the mouse by restriction fragment length variations. Biochem Genet 28:257–266

Wise LS, Green H (1979) Participation of one isozyme of cytosolic glycerolphosphate dehydrogenase in the adipose conversion of 3T3 cells. J Biol Chem 254:273–275

The Molecular Oncology of 12q13–15

B. Rommel, J. Bullerdiek, and W. Schloot

Since the observation of alterations in the same cytogenetical breakpoint region 12q13–15 in a group of solid human tumors, the relationship between the breakpoints and the chromosomal localization of oncogenes or suppressor genes has been of particular interest. In this region, four oncogenes (*sas*, *mdm-2*, *wnt-1*, *gli*) and two putative suppressor genes (*rap1B*, CHOP) have been localized (Fig. 1). The main results obtained concerning the function of these genes during normal development and their implications for tumorigenesis will be discussed in this chapter.

The *gli* gene was first identified in a malignant glioma in which it was found to be amplified more than fifty-fold. Thus, it was thought to provide a significant growth advantage (Kinzler et al. 1987). Later, the *gli* gene was also found to be amplified in some liposarcomas, osteosarcomas, and teratocarcinomas (Kinzler and Vogelstein 1990; Wong et al. 1987; Roberts et al. 1989).

The *gli* gene belongs to the Kruppel gene family, which is characterized by the linkage of the C-terminal histidine from one finger to the next finger through the amino acid sequence HTGEKP (Rosenberg et al. 1986). The *gli* gene encodes a 118-kDa nuclear zink finger protein which has a sequence-specific DNA-binding activity and might influence the transcriptional regulation. This protein is mainly expressed in embryonal carcinoma cell lines and testes, but not in most adult tissues (Kinzler and Vogelstein 1990).

Since the *gli* gene was first localized to 12q13–q14.3 (Kinzler et al. 1987) and later sublocalized to 12q13.3–14.1 (Arheden et al. 1989c), it falls within the translocation breakpoint region frequently found to be involved in lipomas, myxoid liposarcomas, uterine leiomyomas, and pleomorphic adenomas of the parotid gland. Thus, it seemed worthwhile to examine these tumors for amplifications and rearrangements of this gene.

In uterine leiomyomas, neither rearrangements nor an amplification of the *gli* gene were found. Only a low-level expression, similar to that found in normal myometrium, was detected. Therefore, it was concluded that *gli* is not involved in the development of uterine leiomyomas (Arheden et al. 1989b).

University of Bremen, Center of Human Genetics and Genetic Counselling, Leobenerstr. ZHG, 28359 Bremen, Germany

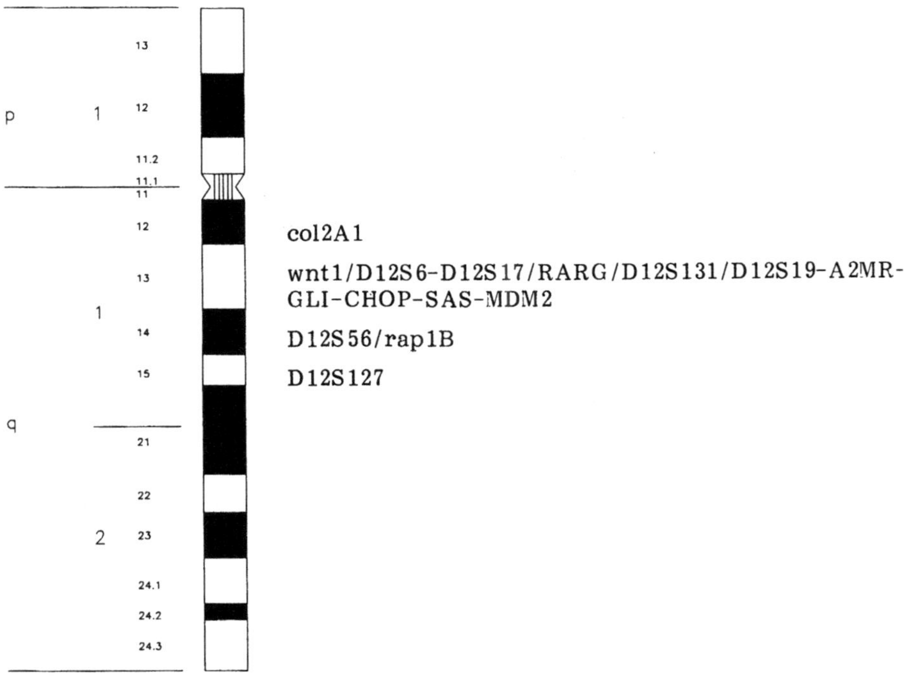

Fig. 1. Localization of proto-oncogenes, putative suppressor genes, and physical markers to chromosomal region 12q13–15 according to the first Chromosome 12 Workshop, Oxford, Sept. 1992

In pleomorphic adenomas of the parotid gland, no rearrangements could be found within 100 kb 3′ and 5′ of the breakpoint region. However, in some of the adenomas, a low-level expression has been detected (B. Vogelstein, personal communication). On the whole, it does not seem probable that the *gli* gene is affected by the structural chromosomal rearrangement of 12q13–15 found in pleomorphic adenomas of the parotid gland.

In myxoid liposarcomas, a translocation t(12;16) is often found. After pulsed field gel electrophoresis analysis, a 600-kb map surrounding the *gli* gene locus was constructed which showed that the chromosomal breakpoint in these tumors is outside that gene. However, altered restriction fragments detectable by a *gli* probe were found in three myxoid liposarcomas (Paulien et al. 1990). According to these authors, the altered fragments are due to a different methylation pattern of this gene which seems to be unique in myxoid liposarcomas.

A gene which has recently been shown to be involved in one of the chromosomal translocations affecting 12q13–15 is CHOP. The CHOP gene product interacts as a dominant– negative inhibitor with transcription factors involved in adipocyte differentiation of fibroblasts and in the growth arrest of terminally differentiated adipocytes (Ron and Habiner 1992). Thus, CHOP appeared to be a candidate gene in fat cell tumorigenesis.

To check for CHOP rearrangements due to the t(12;16) in liposarcomas or the t(3;12) in lipomas, Aman et al. (1992) used a cDNA probe spanning the CHOP coding region. For all nine myxoid liposarcomas tested, one normal and one rearranged allele were found, indicating that the t(12;16) in fact leads to a rearrangement of this gene. It was demonstrated that in most of the tumors tested the breakpoints were in a region containing the first two exons and introns of the gene. In contrast, no CHOP gene rearrangement was found in seven lipomas, two uterine leiomyomas, one hemangiopericytoma, and one chondroma, all having 12q13–15 abnormalities. Also, no rearrangement of CHOP was detected in pleomorphic adenomas characterized by 12q13–15 abnormalities. As for the molecular background of the specific translocation t(12;16)(q13;p11) in myxoid liposarcomas, the translocation leads to the fusion of the CHOP gene with a novel gene FUS (located on chromosome 16), which results in a fusion of the FUS glycine-rich protein with the whole CHOP coding region, corresponding to exons 3 and 4. The FUS sequences are fused to the 5′ side of the CHOP exon 2 noncoding sequence. It is supposed that the FUS–CHOP fusion alters the normal function of CHOP, converting it from a protein involved in growth arrest to one associated with abnormal cellular proliferation (Crozat et al. 1993; Rabbitts et al. 1993).

The proto-oncogene *wnt-1* was first cloned from a mouse tumor as the cellular flanking region of the mouse mammary tumor virus (MMTV; Nusse and Varmus 1982). Subsequently, at least ten *wnt* sequences have been isolated from a variety of species, including zebrafish (Molven et al. 1991), *Drosophila* (Rijsewijk et al. 1987), and humans (van't Veer et al. 1984). The *wnt* genes encode secretory glycoproteins and apparently cause cells to proliferate or differentiate, a phenomenon also related to other secretory proteins such as fibroblast growth factor (FGF) or nerve growth factor (NGF). The *wnt* proteins appear to be associated with the extracellular matrix and the cell surface, too. Up to now, nothing is known about the biological activity of this complex (Nusse and Varmus 1992). Expression of *wnt-1* is normally found during embryonal development (days 8–13) and in the testes of adult mice (Jakobovits et al. 1986). In transgenic mice, expression of *wnt-1* linked to an MMTV long terminal repeat as a transcriptional enhancer had dramatic effects on the growth of mammary glands. Epithelial hyperplasia developed in male and female animals, and after a period of latency mammary carcinomas appeared. Some animals developed adenocarcinomas of the salivary glands as well (Tsukamoto et al. 1988). Tsukamoto et al. (1988) concluded from their experiments that *wnt-1* expression leads to proliferation of epithelial cells in salivary glands and that its activation contributes as an early event to the development of tumors. However, the nude mice experiments clearly show that its expression is per se not sufficient to cause malignant tumors.

Since in humans *wnt-1* was mapped to chromosome 12q13–15 (Turc-Carel et al. 1987b; Arheden et al. 1988b), several tumors with chromosomal breakpoints in that region were checked for alterations of the gene. In lipomas, neither a rearrangement nor an amplification was found (Arheden et al. 1989a). The same applies to liposarcomas (Turc-Carel et al. 1987a), although one case with aberrations of *wnt-1* has been described (Fujii et al. 1989). In uterine leiomyomas, no rearrangements of *wnt-1* were found either (Arheden et al. 1989b). To our knowledge, pleomorphic adenomas of the parotid gland have not yet been tested for alterations in the *wnt-1* gene. Nevertheless, it is probably not affected by the 12q13–15 chromosomal aberration, since by now the *wnt-1* gene is mapped proximal to the chromosomal breakpoint in pleomorphic adenomas (Craig et al. 1992). In summary, in none of the tumors with 12q13–15 aberrations were changes in the *wnt-1* gene found. Up to now, changes in the *wnt-1* gene were found in retinoblastomas. Amplification of *wnt-1* was correlated with higher aggressive potential of the tumor (Arheden et al. 1988a).

Another gene of interest is *rap1B*. Its product belongs to a group of low molecular weight guanosine triphosphate (GTP)-binding proteins which function as molecular switches early in the signal transduction pathways for cell growth and differentiation (Pizon et al. 1988b). These proteins are thought to be guanosine diphosphate (GDP)-ligated in their inactive state, and after an extracellular signal for activation they are converted to the active GTP-ligated state through a nucleotide exchange reaction. They switch back to the inactive state by endogenous GTPase activity, which is enhanced by GTPase-activating proteins (for McCormick see Fischer et al. 1990). The *rap1B* protein shows a 50% homology to the ras proteins. It shows the same C-terminal cysteine needed for membrane anchoring and the same GTP-binding region (Pizon et al. 1988a), which suggests that the proteins coded by the *ras* gene family and by *rap* might interact with the same effector (Pizon et al. 1988b).

The *rap1B* gene has been localized on chromosome 12q14 (Rousseau-Merck et al. 1990). Because of this localization in the breakpoint region of the above-mentioned tumors, DNA of these was screened for rearrangements of the *rap1B*. Neither DNA of lipomas nor uterine leiomyomas showed any differences in fragment length when compared to the germ line fragments of the patients. Only three pleomorphic adenomas of the parotid gland showed additional *Hin*dIII fragments, which probably reflect a restriction fragment polymorphism or a difference in methylation pattern rather than a cancer specific rearrangement (Rommel 1991).

Not only rearrangements, but also another mechanism may play a role in the initiation and/or progression of tumorigenesis: gene amplification, which is often, though not always, cytogenetically manifested by the presence of chromosomes containing homogeneously staining regions (HSR) or paired acentric chromatin bodies termed double minutes. These chromosomal abnormalities often represent amplified oncogenes. One of these genes, which was first found to be amplified in the mouse cell line 3T3-DM, is the *mdm-2* gene (*mdm-1* was first described in that line, too), subsequently localized to mouse chromosome 10. Both *mdm-1* and *mdm-2* were found to be evolutionary conserved sequences. Expression of these genes was detected in a variety of cell types and nontransformed cell lines having myeloid, adrenocortical, and fibroblastic origins, indicating that they might be involved in basic cellular functions (Cahilly-Snyder et al. 1987).

In a later study, nontransformed rat and mouse cells were transfected with altered *mdm-1*, *mdm-2*, and *mdm-3*. Only transfectants overexpressing *mdm-2* acquired tumorigenic properties. These results strongly indicate that *mdm-2* has oncogenic potential and functions in pathways of cellular growth control (Fakharzadeh et al. 1991). Based on the features of its predicted product, which contains a metal-binding region like a DNA-binding protein, an acidic activation domain present in some transcriptional activators, a nuclear localization signal, and a potential protein kinase A recognition site, Fakharzadeh et al. (1991) predicted that *mdm-2* "encodes a potential DNA-binding protein that functions in the modulation of expression of other genes and, when present in excess, interferes with normal constraints on cell growth" (Fakharzadeh et al. 1991). In an attempt to elucidate the function of *mdm-2*, Momand et al. (1992) found that its gene product can regulate *p53* activity (for review on *p53*, see Chen et al. 1990). *mdm-2* is able to inhibit *p53*-mediated transactivation, so the proliferation controlled by *p53* can be abrogated (Momand et al. 1992).

The *mdm-2* gene is highly conserved in different organisms. In humans it has been localized to chromosome 12 (Fakharzadeh et al. 1991) and later sublocalized to 12q13–14 (Oliner et al. 1992). Since this region is often aberrant in human sarcomas, 47 sarcomas of different histologic types were tested by Oliner et al. (1992) for *mdm-2* alterations. Seventeen of these 47 tumors, including seven of 13 liposarcomas, seven of 22 malignant fibrous histiocytomas, and three of 11 osteosarcomas, showed a five- to 50-fold amplification. In five lipomas, one rhabdomyosarcoma, and 74 carcinomas, no amplification was seen (Oliner et al. 1992). Some cell lines were also tested for expression of *mdm-2*. Lines with amplification and one carcinoma line did not show an expression (Oliner et al. 1992). Oliner et al. (1992) concluded from their studies that amplification of *mdm-2* in sarcomas leads to an escape from *p53*-regulated growth control following a mechanism which parallels that for virally induced tumors in which viral oncogene products bind to and functionally inactivate *p53*.

Therefore, in the near future it might be worthwhile to test pleomorphic adenomas of the parotid gland and leiomyomas of the uterus for *mdm-2* alterations, too, in order to find out whether this is a more general change in tumors with 12q13–15 aberrations or whether it is restricted to the abovementioned sarcomas.

References

Aman P, Ron D, Mandahl N, Fioretos T, Heim S, Arheden K, Willén H, Rydholm A, Mitelman F (1992) Rearrangement of the transcription factor gene CHOP in myxoid liposarcomas with t(12;16)(q13;p11). Genes Chrom Cancer 5:278–286

Arheden K, Tommerup N, Mandahl N, Heim S, Winther J, Jensen OA, Prause JU, Mitelman F (1988a) Amplification of the human putative oncogene int1 in primary retinoblastoma tumors. Cytogenet Cell Genet 48:174–177

Arheden K, Mandahl N, Strömbeck B, Isaksson M, Mitelman F (1988b) Chromosome localization of the human oncogene int1 to 12q13 by in situ hybridization. Cytogenet Cell Genet 47:86–87

Arheden K, Mandahl N, Heim S, Mitelman F (1989a) The int1 oncogene is not rearranged or amplified in lipomas with structural chromosomal abnormalities of 12q13–15. Cancer Genet Cytogenet 42:143–146

Arheden K, Nilbert M, Heim S, Mandahl N, Mitelman F (1989b) No amplification or rearrangement of INT1, GLI, or COL2A1 in uterine leiomyomas with t(12;14)(q14–15; q23–24). Cancer Genet Cytogenet 39:195–201

Arheden K, Ronne M, Mandahl N, Heim S, Kinzler KW, Vogelstein B, Mitelman F (1989c) In situ hybridization localizes the human putative oncogene gli to chromosome subbands 12q13.3–14.1. Hum Genet 82:1–2

Cahilly-Snyder L, Yang-Feng T, Francke U, George DL (1987) Molecular analysis and chromosomal mapping of amplified genes isolated from a transformed mouse 3T3 cell line. Somat Cell Mol Genet 13:235–244

Chen P-L, Chen Y, Bookstein R, Lee WH (1990) Genetic mechanisms of tumor suppression by the human p53 gene. Science 250:1576–1580

Craig I, Gemmill R, Kucherlapati R, Craig S (1992) Report of the first international workshop on human chromosome 12 mapping. Cytogenet Cell Genet 61:243–262

Crozat A, Aman P, Mandahl N, Ron D (1993) Fusion of CHOP to a novel RNA-binding protein in human myxoid liposarcoma. Nature 363:640–644

Fakharzadeh SS, Trusko SP, George DL (1991) Tumorigenic potential associated with enhanced expression of a gene that is amplified in a mouse tumor cell line. EMBO J 10:1565–1569

Fischer TH, Gatling MN, Lacal J-C, White GC (1990) Rap1B, a cAMP-dependent protein kinase substrate, associates with the platelet cytoskeleton. J Biol Chem 265:19405–19408

Fujii Y, Matui Y, Nakagawa Y, Hongo T, Igarashi Y, Yamada M, Nakagome Y, Tobayama S (1989) Translocation t(12;16)(q13;p11) in myxoid liposarcoma of a child and implication of the human int-1 gene in tumorigenesis. Jpn J Cancer Res 80:958–962

Jakobovits A, Shackleford GM, Varmus HE, Martin GR (1986) Two proto-oncogenes implicated in mammary carcinogenesis, int-1 and int-2, are independently regulated during mouse development. Proc Natl Acad Sci USA 83:7806–7810

Kinzler KW, Vogelstein B (1990) The gli gene encodes a nuclear protein which binds specific sequences in the human genome. Mol Cell Biol 10:634–642

Kinzler KW, Bigner SH, Bigner DD, Trent JM, Law ML, O'Brien SJ, Wong AJ, Vogelstein B (1987) Identification of an amplified, highly expressed gene in a human glioma. Science 236:70–73

Molven A, Njolstad P, Fjose A (1991) Genomic structure and restricted neural expression of the zebrafish wnt-1 (int-1) gene. EMBO J 10:799-807

Momand J, Zambetti GP, Olson DC, George D, Levine AJ (1992) The mdm-2 oncogene product forms a complex with the p53 protein and inhibits p53-mediated transactivation. Cell 69:1237-1245

Nusse R, Varmus HE (1982) Many tumors induced by the mouse mammary tumor virus contain a provirus integrated in the same region of the host genome. Cell 31:99–109

Nusse R, Varmus HE (1992) Wnt genes. Cell 69:1073–1987

Oliner JD, Kinzler KW, Meltzer PS, George DL, Vogelstein B (1992) Amplification of a gene encoding a p53-associated protein in human sarcomas. Nature 358:80–86

Paulien S, Turc-Carel C, Dal Cin P, Jani-Sait S, Sreekantaiah C, Leon GS, Vogelstein B, Kinzler KW, Sandberg AA, Gemmill RM (1990) Myxoid liposarcoma with t(12;16)(q13;p11) contains site specific alterations in methylation patterns surrounding a zinc-finger gene mapped to the breakpoint region on chromosome 12. Cancer Res 50:7902–7908

Pizon V, Chardin P, Lerosey I, Olofsson B, Tavitian A (1988a) Human cDNAs rapl and rap2 homologous to the Drosophila gene Dras3 encode proteins closely related to ras in the effector region. Oncogene 3:201–204

Pizon V, Lerosey I, Chardin P, Tavitian A (1988b) Nucleotide sequence of a human cDNA encoding a ras-related protein (raplB). Nucleic Acids Res 16:7719

Rabbitts TH, Forster A, Larson R, Nathan P (1993) Fusion of the dominant negative transcription regulator CHOP with a novel gene FUS by translocation t(12;16) in malignant liposarcoma. Nature Genet 4:175–180

Rijsewijk F, Schuermann M, Wagenaar E, Parren P, Weigel D, Nusse R (1987) The Drosophila homolog of the mouse mammary oncogene int-1 is identical to the segment polarity gene wingless. Cell 50:649–657

Roberts WM, Douglass EC, Peiper SC, Houghton PJ, Look AT (1989) Amplification of the gli gene in childhood sarcomas. Cancer Res 49:5407–5413

Rommel B (1991) Molekulargenetische Untersuchungen zum Einfluß bzw. zur Aktivierung von Proto-Onkogenen als Folge instabiler Chromatinkonfigurationen bei benignen Tumoren. Thesis, University of Bremen

Ron D, Habiner JF (1992) CHOP, a novel developmentally regulated nuclear protein that dimerizes with transcription factors C/EBP and LAP and functions as a dominant negative inhibitor of gene transcription. Genes Dev 6:439–453

Rosenberg UB, Schroder C, Preiss A, Kienlin A, Cote S, Riede I, Jackle H (1986) Structural homology of the product of the Drosophila Kruppel gene with Xenopus transcription factor IIIA. Nature 319:336–339

Rousseau-Merck MF, Pizon V, Tavitian A, Berger R (1990) Chromosome mapping of the human ras-related raplA, raplB, and rap2 genes to chromosomes 1p12–13, 12q14, and 13q34, respectively. Cytogenet Cell Genet 52:2–4

Tsukamoto AS, Grosschedl R, Guzman RC, Parslow T, Varmus HE (1988) Expression of the int-1 gene in transgenic mice is associated with mammary gland hyperplasia and adenocarcinomas in male and female mice. Cell 55:619–625

Turc-Carel C, Dal Cin P, Limon J, Gibas Z, Kakati S, Pietrzak E, Parry D, Li F, Sandberg AA (1987a) Cytogenetic and molecular studies of adipose tissue tumors. Cancer Genet Cytogenet 28:33

Turc-Carel C, Pietrzak E, Kakati S, Kinniburgh AJ, Sandberg AA (1987b) The human int-1 gene is located at chromosome region 12q12–12q13 and is not rearranged in myxoid liposarcoma with t(12;16)(q13;p11). Oncogene Res 1:397–405

Van'T Veer L, van Kessel AG, van Heerikhuizen H, van Ooyen A, Nusse R (1984) Molecular cloning and chromosomal assignment of the human homolog of int-1, a mouse gene implicated in mammary tumorigenesis. Mol Cell Biol 4:2532–2534

Wong AJ, Bigner SH, Bigner DD, Kinzler KW, Hamilton SR, Vogelstein B (1987) Increased expression of the epidermal growth factor receptor gene in malignant gliomas is invariably associated with gene amplification. Proc Natl Acad Sci USA 84:6899–6903

Gene Analysis – Methods and Facts

Cell Lines from Tumors Showing 12q13–15 Aberrations

B. Kazmierczak[1], S. Bartnitzke[1], P. Dal Cin[2], R. Vanni[3], and J. Bullerdiek[1]

1 Introduction

Cells of human benign neoplasms such as pleomorphic adenomas of the parotid gland (PSA), uterine leiomyomas, lipomas, and even malignant myxoid liposarcomas stop proliferation after 3–4 in vitro passages, not allowing sufficient material for molecular analyses by in vitro preparation to be obtained (for PSA, see Stern et al. 1990). One alternative to overcome the problem of replicative senescence (Hayflick and Moorhead 1961) of tumor cells in culture is the establishment of virus-transformed cell lines. Several viruses with oncogenic potential, e.g., viruses belonging to the group of papovavirus, adenovirus, or herpesvirus are well known to induce transformation of cells in vitro. For example, the Epstein-Barr virus (EBV) immortalizes human primary B cells in vitro. This process may be divided into three steps: (1) B-cell activation by binding of the virus to its receptor, (2) sequential expression of several EBV genes, and (3) indefinite proliferation leading to lymphoblastoid cell lines. For the establishment and maintenance of the immortal state of the host cells, continuous expression of a subset of the EBV viral genes is necessary.

The successful cell transformation in vitro by a virus largely depends on the origin of the cell. Thus, the condition of cells and viruses is extremely important in designing transformation experiments.

For transformation of cells originating from pleomorphic adenomas of the salivary gland, uterine leiomyomas, lipomas, and myxoid liposarcomas, we decided to use a DNA construct containing the early region of the simian virus 40 (SV40) coding for small and large T-antigens (Fig. 1).

SV40 is a small DNA tumor virus which can transform host cells or replicate and lytically infect the cells. Cellular transformation and viral replication are both initiated by an SV40 early gene product, large tumor

[1] Center for Human Genetics and Genetic Counselling, University of Bremen, ZHG, Leobenerstr., 28359 Bremen, Germany
[2] Center for Human Genetics, U.Z. Gasthuisberg herestraat 49, 3000 Leuven, Belgium
[3] Institute for General Biology, Faculty of Medicine and Surgery, University of Cagliary, Via Ospedale 119, 09124 Cagliary, Italy

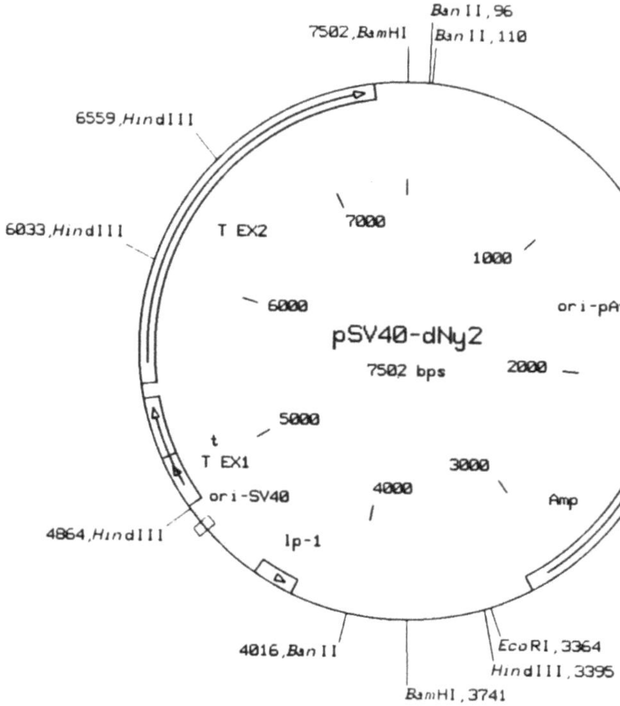

Fig. 1. The organization of the plasmid pSV40dN. For the Construction of pSV40dN, the SV40 BanII fragment (from nucleotide 776 to position 2285, late region) was deleted and the remaining sequences of the SV40 genome were cloned into the *Bam*HI site of plasmid pAT153. The relative positions of the replication origin and the coding sequences for small and large T-antigens are indicated. SV40, simian virus 40

antigen (T-antigen). One activity of T-antigen believed to contribute to transformation is the formation of complexes with the cellular protein p53 (Lane and Crawford 1979). SV40-transformed cells are generally characterized by nuclear localization of T-antigen and p53, increased mitotic rate, anchorage-independent cell growth, and an increased life span before the cells enter crisis. Crisis is characterized by a state of culture decline in which the cells do not arrest in G_0 showing a high number of dying cells. Crisis is further marked by abnormal mitotic figures, a decrease in proliferative capacity, detachment from the surface, and formation of giant or multinucleated cells. Recovering of cells from crisis is a rare event which subsequently leads to immortalized cell lines.

SV40-mediated transformation of both human mesenchymal and epithelial cells has been studied for more than 25 years. The establishment of numerous cell lines originating from human fibroblasts (Nichol 1983), liver cells (Buitenwerf et al. 1982), or uroepithelial cells (Christian et al. 1987) has since been described. The transformation of slowly proliferating human solid tumors with specific chromosomal aberrations has, nevertheless, only occasionally been reported before. The disadvantage in transforming cells of tumors such as pleomorphic adenomas, lipomas, leiomyomas, and myxoid liposarcomas is their low proliferative capacity, which makes integration and expression of viral sequences very inefficient.

All these tumors have subsets showing aberrations of chromosome 12 with the breakpoint region in 12q13–15. Virus-transformed cell lines containing rearranged chromosomes are a very suitable tool to elucidate the mechanisms underlying the genesis of these tumors. In order to obtain sufficient cellular material for molecular investigations, we developed a transfection method to establish cell lines of all four tumors (Kazmierczak et al. 1990). Because of their high proliferative capacity, transformed cells can be used for methods such as high-resolution banding, in situ hybridization, and DNA extraction. Furthermore, somatic cell hybrids between cells of an SV40-transformed liposarcoma cell line with a translocation t(12;16) and trisomy 8 and mouse A9 cells have been established (Schoenmakers et al. 1993). The cell hybrids contained der(16) as the only human derivative chromosome and were lacking the normal human chromosomes 12 and 16.

Besides the possibility of getting sufficient cellular material for molecular investigations of the breakpoint region, the transfection assay is also helpful in developing an in vitro model for malignant transformation of pleomorphic adenomas (Kazmierczak et al. 1992a; Kazmierczak 1993): for the pleomorphic adenomas, it is well known that carcinomas can arise within the adenomas after insufficient surgical removal or prolonged growth periods. Up to now, we know very little about the genetic changes responsible for these phenomena. However, the similar characteristics exhibited by virus-transformed and tumor cells are the reason that in vitro cellular transformation by viruses is a widely employed model for studying the molecular events leading to neoplasia. Expression of large T-antigens is capable of transforming cells, but for the process of immortalization and tumorigenicity in nude mice, additional mutational changes are necessary. Large T-antigens are well known to induce secondary chromosomal aberrations. Analysis of these secondary changes gave rise to an in vitro model for the progression from benign pleomorphic adenomas to undifferentiated carcinomas in nude mice. For these studies, immortalized cells of pleomorphic adenomas were injected into nude mice.

2 Results of Transfection Experiments

In order to obtain a huge panel of transformed cell lines of all the different types of tumors with 12q13–15 aberrations, we transfected the cells of the primary cultures with the SV40 plasmid pSV40-dN. This plasmid contained the early region of the SV40 genome coding for small and large T-antigens and was cloned in the BamHI site of the vector pAT 153. The SV40 genome was deleted from positions 776–2258 bp (BanII fragment; Fig. 1).

On average, the first clones of transformed cells usually appeared after 34 days. The transformed cells showed all typical SV40 transformation patterns (Shein and Enders 1962), including altered morphology, growth in soft agar, loss of contact inhibition, and growth in medium containing low serum concentrations and also had an increased mitotic rate. The transformed cells were visible as rapidly growing foci in the monolayer and could easily be recognized by their altered morphology and their increased mitotic rate. The integration of the viral sequences was routinely checked by Southern blot analysis after digestion with a restriction enzyme that had only one cut site in the plasmid. Expression of large T-antigens was shown by indirect immunofluorescence using rabbit anti-T antibodies and fluorescein isothiocyanate (FITC)-labeled anti-rabbit antibodies. All transformed cells showed a bright nuclear T-antigen staining (Fig. 2).

Since it is well known that in most cases the primary tumor cultures are contaminated with normal diploid cells of the surrounding tissue, the cells were karyotyped directly after transformation. All transformed cell lines had retained their initial translocation chromosomes resulting from the structural alterations of the primary tumors, thus proving that in point of fact tumor cells have been transformed.

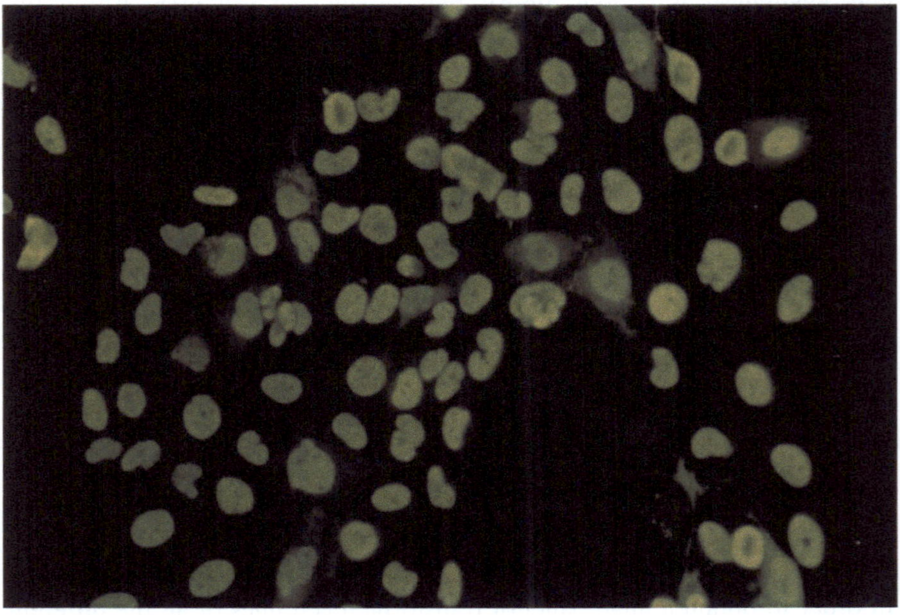

Fig. 2. Simian virus 40-transformed cells of a pleomorphic adenoma. Nuclear localization of large T-antigen was shown by indirect immunofluorescence. For detection, T-antigen antibodies and fluorescein isothiocyanate-labeled anti-antibodies were used

The transformed cells also had an extended life span compared to the nontransformed control cells of the primary cultures. For PSA, the average number of passages reached by cells from the primary tumor material was 3.7 (Stern et al. 1990), whereas the transformed cells usually showed an enhanced growth capacity up to the 13th–16th passage (precrisis stage). After this precrisis stage, the cells entered crisis, which was characterized by reduced proliferative capacity, loss of attachment to the bottom of the culture flasks, and formation of multinuclear giant cells, as also reported for other types of human cells transformed by SV40 (Girardi et al. 1965). Crisis lasts for between 3–4 weeks and 3 months and recovering from crisis is a very rare event. Cells having escaped from crisis are then immortalized (postcrisis). In all, we transfected 66 primary cultures of different types of tumors showing different primary chromosomal aberrations (Table 1).

As well as tumors with translocations involving the breakpoint region in 12q13–15 such as pleomorphic adenomas (Kazmierczak et al. 1990, 1992b), uterine leiomyomas (Stern et al. 1991), lipomas (Belge et al. 1992), and myxoid liposarcomas (Schoenmakers et al. 1993), we also tried to transform pleomorphic adenomas with aberrations of chromosome 8 with the specific breakpoint in 8q12. Whereas the transformation efficiency was very high for tumors having primary aberrations on chromosome 12 such as leiomyomas (8/9), pleomorphic adenomas (7/7), and lipomas (5/6), we detected a significantly lower transformation efficiency for pleomorphic adenomas showing aberrations of 8q12 (Kazmierczak et al. 1992b).

In addition, the immortalization rate was significantly higher for adenomas with a primary aberration of 12q13–15 (3/7) than for adenomas

Table 1. Results of transfection experiments

Type of tumor and breakpoint	Transfected tumors	Contaminated	Transformed	Not transformed
PSA 8q12	26	4	4	18
PSA 12q13–15	7	–	7	–
PSA others	5	–	3	2
Leiomyoma 12q13–15	11	2	8	1
Leiomyoma 7q-	6	–	–	6
Lipoma 12q13–15	6	–	5	1
Liposarcoma 12q13–15	5	–	1	4
Total	66	6	28	32

PSA, pleomorphic adenomas of the parotid gland.

having aberrations of 8q12 (2/18) (Kazmierczak et al. 1992). The relatively low frequency of immortalization was in agreement with results reported by others for human Fibroblasts (Chang 1986; Shay and Wright 1989). Immortalized cell lines have also been established from a variety of uterine leiomyomas with 12q13–15 aberrations, one lipoma, and one myxoid liposarcoma. Different susceptibility of cells to SV40 transformation has been discussed earlier for different human fibroblast cell lines with predominant intrinsic genetic changes (Kersey et al. 1972; Payne and Schmickel 1971; Todaro et al. 1966). In all the tumors investigated in our series, the susceptibility seems to depend on the primary aberrations observed. In these cases, there seems to be a stronger cooperative effect between the hypothesized, activated proto-oncogene-like sequences in 12q13–15 and large T-antigen than with the oncogenic sequences located at 8q12.

As far as the pleomorphic adenomas are concerned, there seem to be at least two different mechanisms involved in the initiation of these tumors by activating different proto-oncogene-like sequences located in the different breakpoint regions 8q12 and 12q13–15.

By transfecting different types of tumors with primary chromosomal changes, we not only can provide ample cellular material for molecular investigations, but also gain more insight into the biological properties of the hypothesized proto-oncogene-like sequences located at the different breakpoint regions by way of a complementation assay. For example, it can be hypothesized that in the case of pleomorphic adenomas, genes located at 12q13–15 seem to have a higher malignant potential than sequences located at 8q12. It is well known that pleomorphic adenomas can develop malignant carcinomas from benign adenomas in vivo. This process is possibly related to the type of primary aberration in the benign tumors. For the situation in vivo, it could be speculated that carcinomas then predominantly rise out of adenoma cells showing aberrations of the breakpoint region 12q13–15.

Extensive karyotype analyses were performed of pre- and postcrisis cells. Apart from checking the transformed cells to discover whether they had retained their initial translocation chromosomes, a second reason was the monitoring of the karyotypic instability caused by SV40 large T-antigen expression. Several SV40-transformed cell lines have previously been analyzed for their chromosomal patterns, and most of the authors detected a gross disturbance of the karyotype soon after expression of the viral oncogenes and prior to crisis (Wolman et al. 1964; Legrand et al. 1991; Christian et al. 1987; Ray et al. 1990). From these observations it was concluded that viral oncogenes might directly or indirectly alter the integrity and stability of the cellular genome (Stewart and Bacchetti 1991). Surprisingly, in our series of experiments most of the transformed cell lines remained karyotypically relatively stable (Kazmierczak et al. 1990; Stern et al. 1991; Schoenmakers et al. 1993) during precrisis, generating only a few sporadic secondary marker chromosomes, which made them very suitable for molecular investigations of the breakpoint region (Fig. 3a). The gross disturbance of the karyotype

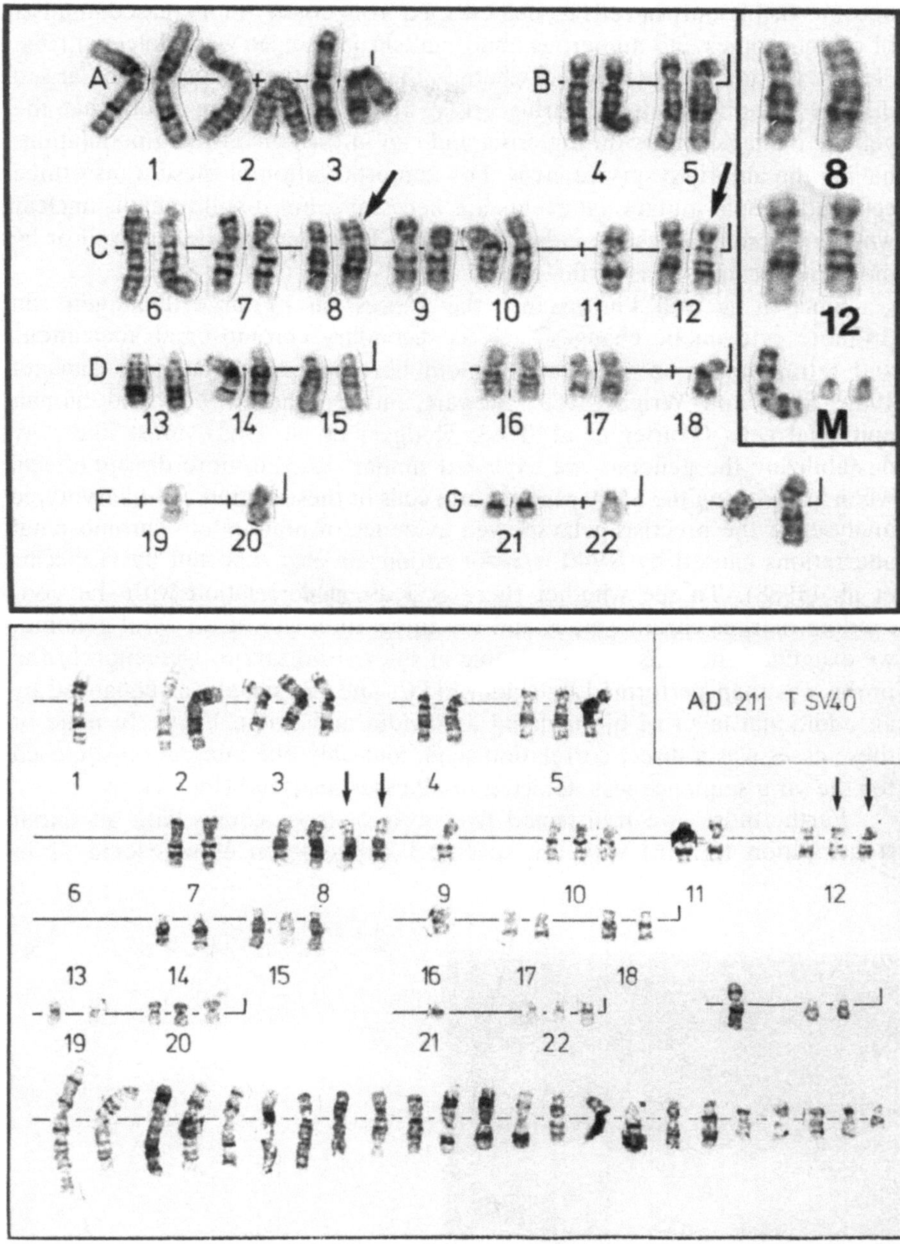

Fig. 3. **a** Representative karyotype of the simian virus 40 (SV40)-transformed pleomorphic adenoma 211 in the precrisis stage. The primary translocation chromosomes are indicated by *arrows*. *Inlet*, enlarged normal and aberrant chromosomes 8 and 12 of an other metaphase and sporadic aberrant chromosomes caused by large T-antigen expression. *A-G*, chromosome groups; *M*, marker chromosomes. **b** Representative karyotype of the SV40-transformed pleomorphic adenoma 211 (*AD 211*) in the postcrisis stage. The primary translocation chromosomes are indicated by *arrows*. *Bottom line*, abnormal chromosomes caused by large T-antigens

became visible only in cells having escaped from crisis. An increased number of chromosomes and numerous abnormal chromosomes were detected (Fig. 3b). It cannot be determined whether these additional aberrations arised directly prior to crisis or during crisis, but it seems more likely that the genetic damage occurs during crisis and can in turn give rise to populations having the ability to survive crisis. For immortalization of these transformed cells, additional mutational events are necessary, but it still remains unclear whether these changes are induced by large T-antigen expression itself or by independent mutations of the host cells.

Since it is well known that the expression of large T-antigen can promote cytogenetic changes such as secondary chromosomal aberrations and tetraploidization in human diploid fibroblasts (Naiman and Canaani 1989; Shay and Wright 1989; Stewart and Bacchetti 1991) and human epithelial cells (Poirier et al. 1988; Rodgers et al. 1983), most likely by destabilizing the genome, we expected similar or even more drastic effects when introducing the SV40 plasmid into cells of these tumors. But karyotypic analyses of the precrisis cells showed evidence of nonrandom chromosomal aberrations caused by SV40 transformation, as also reported by Hoffschir et al. (1988). To see whether there is a direct correlation with the nonrandom marker chromosomes and the integration site of the viral genome, we examined the cells by nonisotopic in situ hybridization. Detection of the probe was then performed by avidin FITC, and the signal was enhanced by an additional layer of biotinylated antiavidin and avidin FITC. In none of these cases was a direct correlation seen, and only one integration site each for the viral sequence was detected in all cases analyzed (Fig. 4).

Furthermore, we transfected five myxoid liposarcomas with an initial translocation t(12;16) with the specific breakpoint on chromosome 12 in

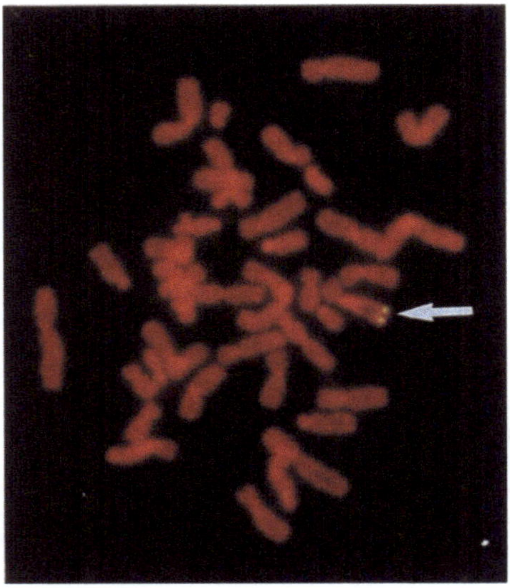

Fig. 4. Metaphase spread of a simian virus 40 (SV40)-transformed pleomorphic adenoma showing one integration site for the viral sequences (*arrow*). Detection of the biotinylated probe was performed by avidin fluorescein isothiocyanate and counterstaining of the chromosomes by propidium iodide

12q13–15. All of them were transformed, but only one liposarcoma showed the primary translocation chromosomes. The others had a normal diploid karyotype, showing that fibroblasts of the surrounding tumor tissue had been transformed. The transformed cells of the myxoid liposarcoma showed the t(12;16) translocation and an additional trisomy 8. The cells have not entered crisis up to now (110th passage, 1:2 splitting ratio), still showing a high growth potential. The cells are T-antigen positive and have a remarkable stable karyotype. The last karyotype analysis was performed in the 60th passage, showing no further clonal karyotypic alterations. More myxoid liposarcomas have to be investigated to see whether this is a typical feature of SV40-transformed liposarcomas. This shows that cytogenetic alterations are not necessarily associated with the expression of SV40 large T-antigens. There are two possible explanations for this phenomenon:

1. Expression of SV40 large T-antigens itself does not lead to immortalization of transformed cells. Whereas the benign tumors with aberrations of the region 12q13–15 immortalized only by additional mutational events, the myxoid liposarcoma was immortalized without any further aberrations.

There are different genetic mechanisms underlying the translocations observed in the liposarcomas and the benign tumors, allowing us to distinguish between benign and malignant tumors. Only in case of liposarcoma could the additional expression of the viral oncogene then contribute to a fully transformed phenotype not dependent upon further mutations of the cellular genome for immortalization.

2. These liposarcoma cells had an additional trisomy 8 in the karyotypes of the primary cultures. Trisomy 8 has also been observed in adenomas of the colon (Reichmann et al. 1982), pleomorphic adenomas of the salivary glands (Mark et al. 1983), breast cancer (Bullerdiek et al. 1993), and leukemias (Sandberg 1986). In both chronic and acute myeloid leukemias, trisomy 8 is even known to correlate with a poor prognosis. Uncontrolled expression of genes located on chromosome 8 such as for example *MOS*, *MYC*, or *GLI4* may enhance the malignant character of the cells. It is possible that cooperational effects of the activated sequences in 12q13–15, the viral oncogene large T-antigen, and a third event initiated by the trisomy 8 exist, so that further mutational alternations of the cellular genome are unnecessary for shifting the cells directly to the immortalized state.

3 Tumorigenicity in Nude Mice

The transformed cells of the myxoid liposarcoma were also tested for tumorigenicity in nude mice. Nude mice were injected subcutaneously with 10^7 cells, and all mice showed tumors after a period of 3 months. The tumor cells grown in the mice were recultivated again after tumor resection.

Transformed cells of two pleomorphic adenomas with primary translocations on 12q13–15 or 8q12, respectively, were also checked for tumorigenicity.

Tumorigenicity was only positive for SV40-transformed cells with an initial translocation on chromosome 12, whereas those with an aberration on 8q12 were not capable of forming tumors in nude mice. When SV40-transformed cells with an initial 12q13–15 aberration were subcutaneously injected into nude mice, slowly growing nodules appeared at the injection site after 7 days and disappeared again. After 74 days, very agressively growing metastasis of undifferentiated carcinomas developed. Cells of these tumors were also recultivated in vitro and checked for the expression of large T-antigens, integration site of the viral DNA, and the presence of the initial translocation chromosomes. All the cells were T-antigen positive and had the same integration site for the viral sequences as before injection, and the translocation chromosomes were also still present. Tumorigenicity of cells transformed by SV40 alone has only rarely been observed for human epithelial cells (Kahn et al. 1983). However, it is not yet clear why SV40-transformed human epithelial cells are usually nontumorigenic in nude mice, while SV40-transformed rodent epithelial cells, such as rat hepatocytes (Woodworth et al. 1988), are often strongly tumorigenic (Freedman and Shin 1978). One explanation may be that cell lines immortalized by viral genes might therefore represent an intermediate stage between normal and neoplastic. The hypothesis that these cells are incompletely transformed is strongly supported by the fact that some SV40-transformed cell lines can be rendered tumorigenic by activated *ras* oncogenes (Clark et al. 1988) or chemical carcinogens such as 3-methylcholanthrene (Reznikoff et al. 1988).

Referring to our model of SV40 transformation of human tumors showing specific chromosomal aberrations, this again possibly reflects the importance of the type of primary aberration prior to exposure of the viral oncogene, akin to the higher transformation efficiency of pleomorphic adenoma cells showing the 12q13–15 aberration compared to adenoma cells with 8q12 aberrations, as already mentioned (Kazmierczak et al. 1992b). For surviving crisis, additional cellular events seem to be necessary for the cells and the cooperation between the oncogenic sequence at the breakpoint and large T-antigens plus additional mutational alterations is possibly strong enough to give rise to tumorigenic cells within this culture. It ought to be investigated whether pleomorphic adenomas with aberrations other than on 12q13–15 become tumorigenic by additional exposure to chemical carcinogens or other activated oncogenes such as the *ras* oncogene.

4 Conclusion

By establishing numerous cell lines of all tumors with 12q13–15 aberrations discussed above, ample cellular material is now available for molecular studies of the breakpoint region. Furthermore, these cell lines will be necessary for defining the distinct mechanisms involved in tumorigenesis when the involved genes are characterized. For the pleomorphic adenomas, a correlation between the primary translocation and the transformation efficiency has been shown among adenomas with either chromosome 8 or chromosome 12 aberrations. This might give us an idea of the different malignant potentials of the proto-oncogenes thought to be involved in tumorigenesis. In addition, we were able to provide the first model for malignant transformation of pleomorphic adenomas by immortalization of adenoma cells with the viral oncogene large T-antigen and subsequent injection of these cells into nude mice, leading to an undifferentiated carcinoma similar to the situation described in vivo.

5 Transfection Method

Because all these tumors are slowly proliferating in vitro and a high proliferation rate is necessary for incorporation of the viral DNA into the host genome, we (Kazmierczak et al. 1990) modified the calcium phosphate coprecipitation method by Graham and van der Eb (1973) for these tumor cultures. The highest rate of transfection was reached when subconfluent monolayers of primary cultures showing a maximal rate of mitosis were used. Prior to transfection, the cells were washed once with Mg^{2+}- and Ca^{2+}-free phosphate-buffered saline. The transfection mixture was prepared by gently mixing the transfection buffer (0.7 m*M* Na_2HPO_4, 21 m*M* Hepes, 0.145 *M* NaCl, pH 7.0) with 12.5–30 μg plasmid DNA, 5–10 μg total human DNA, and 30 μl 2.5 *M* $CaCl_2$. (All solutions were filter-sterilized and kept in sterile plastic tubes prior to use.) The mixture was incubated at 20°C for 20 min and then added to the cells for 45–60 min at 20°C. After that, culture medium was added and the cells were further incubated at 37°C (5% CO_2) for about 5 h. To avoid destruction of the treated cells, they were examined every 15 min and the medium was changed after 5 h.

After the transfection procedure the cells were treated following routine methods and examined twice a week.

Acknowledgements. This work was supported by the Deutsche Forschungsgemeinschaft, the Tönjes-Vagt-Foundation, and the EC Medical and Health Research Programme.

References

Belge G, Kazmierczak B, Meyer-Bolte K, Bartnitzke S, Bullerdiek J (1992) Expression of SV40 T-antigen in lipoma cells with a chromosomal translocation t(3;12) is not sufficient for direct immortalization. Cell Biol Int Rep 16:339–347

Bullerdiek J, Leuschner E, Taquia E, Bonk U, Bartnitzke S (1993) Trisomy 8 as a recurrent clonal abnormality in breast cancer? Cancer Genet Cytogenet (in press)
Buitenwerf J, deJang W, vanStrien A, van der Noordaa J (1982) Characterization of SV40 induced transformed human liver cells before and after passage through crisis. Intervirology 17:222–227
Chang SE (1986) In vitro transformation of human epithelial cells. Biochim Biophys Acta 823:161–194
Christian BJ, Loretz LJ, Oberley TD, Reznikoff CA (1987) Characterization of human uroepithelial cells immortalized in vitro by Siman virus 40. Cancer Res 47:6066–6073
Clark R, Stampfer MR, Miller R (1988) Transformation of human mammary epithelial cells by oncogenic retroviruses. Cancer Res 48:4689
Freedman VH, Shin S (1978) Use of nude mice for studies on the tumorigenicity of animal cells. In: Fogh J, Giovanella B (eds) The nude mouse in experimental and clinical research. Academic, New York, p 353
Girardi AJ, Jensen FC, Koprowski H (1965) SV40 induced transformation of human diploid cells: crisis and recovery. J Cell Comp Physiol 65:69–84
Graham FL, Van der Eb AJ (1973) A new technique for the assay of infectivity of human adenovirus 5 DNA. Virology 52:456–467
Hayflick L, Moorhead PS (1961) The serial cultivation of human diploid cell strains. Exp Cell Res 25:585–621
Hoffschir F, Ricoul M, Dutrillaux B (1988) SV40-transformed human fibroblasts exhibit a characteristic chromosomal pattern. Cytogenet Cell Genet 49:264–268
Kahn P, Topp WC, Shin SI (1983) Tumorigenicity of SV40-transformed human and monkey cells in immunodeficient mice. Virology 126:348
Kazmierczak B (1993) SV40 transformation as an in vitro model for malignant transformation. Thesis, University of Bremen
Kazmierczak B, Bartnitzke S, Hartl M, Bullerdiek J (1990) In vitro transformation by the SV40 early region of cells from a human benign salivary gland tumor with a 12q13–15 rearrangement. Cytogenet Cell Genet 53:37–39
Kazmierczak B, Meyer-Bolte K, Bartnitzke S, Bullerdiek J (1992a) Immortalization of human benign epithelial tumor cells as a model for malignant transformation. J Cancer Res Clin Oncol 118 [Suppl]:R152
Kazmierczak B, Thode B, Bartnitzke S, Bullerdiek J, Schloot W (1992b) Pleomorphic adenoma cells vary in their susceptibility to SV40 transformation depending on the initial karyotype. Genes Chrom Cancer 5:35–39
Kersey JH, Gatti RA, Good RA, Aaronson SA, Todaro GJ (1972) Susceptibility of cells from patients with primary immunodeficiency diseases to transformation by simian virus 40. Proc Natl Acad Sci USA 69:980–982
Lane DP, Crawford LV (1979) T antigen is bound to a host protein in SV40-transformed cells. Nature 278:261–263
Legrand A, Greenspan P, Nagpal ML, Nachtigal SA, Nachtigal M (1991) Characterization of human vascular smooth muscle cells transformed by the early genetic region of SV40 virus. Am J Pathol 139:629–639
Mark J, Dahlenfors R, Ekedahl C (1983) Cytogenetics of the human mixed salivary gland tumour. Hereditas 99:115–129
Naiman T, Canaani D (1989) A hypodiploid karyotype found in immortal human cells selected from a wide spectrum of posttransformation chromosomal complements. Cancer Genet Cytogenet 40:65–71
Nichol VJ (1983) Characterization of a new human diploid cell line. In Vitro 19:797–804
Payne FE, Schmickel RD (1971) Susceptibility of trisomic and of triploid human fibroblasts to simian virus 40 (SV40). Nature New Biol 230:190
Poirier V, Taylor SJ, Brown KW, Shaw APW, Maitland NJ (1988) SV40 transfection of human kidney epithelial cells and stability of chromosome 11. Int J Cancer 42:887–894

Ray FA, Peabody DS, Cooper JL, Cram LS, Kraemer PM (1990) SV40 T antigen alone drives karyotype instability that precedes neoplastic transformation of human diploid fibroblasts. J Cell Biochem 42:13–31

Reichmann A, Martin P, Levin B (1982) Karyotypic findings in a colonic villous adenoma. Cancer Genet Cytogenet 7:51–57

Reznikoff CA, Loretz LJ, Christian BJ, Wu S-Q, Meisner LF (1988) Neoplastic transformation of SV40-immortalized human urinary tract epithelial cells by in vitro exposure to 3-methylcholanthrene. Carcinogenesis 9:1427–1436

Rodgers CS, Hill SM, Hulten MA, Chang SE, Keen J, Taylor-Papadimitriou J (1983) Cytogenetic analysis of SV40 transformed human breast epithelial cells. Cancer Genet Cytogenet 8:213–221

Sandberg AA (1986) The chromosomes in human leukemia. Semin Hematol 23:201–217

Schoenmakers EFPM, Kazmierczak B, Kools PFJ, Mols R, Dal Cin P, Bullerdiek J, Van den Berghe H, Van de Ven WJM (1993) Development and characterization of cell lines from a myxoid liposarcoma with t(12;16)(q13;p11.2) and trisomy 8. Cytogenet Cell Genet (in press)

Shay JW, Wright WE (1989) Quantitation of the frequency of immortalization of normal human diploid fibroblasts by SV40 large T-antigen. Exp Cell Res 184:109–118

Shein HM, Enders JF (1962) Multiplication and cytopathogenicity of simian vacuolating virus 40 in cultures of human tissues. Proc Soc Exp Biol Med 109:495–500

Stern C, Meyer K, Bartnitzke S, Schloot W, Bullerdiek J (1990) Pleomorphic adenomas with unbalanced chromosomal abnormalities have an increased in vitro lifetime. Cancer Genet Cytogenet 46:55–63

Stern C, Kazmierczak B, Thode B, Rommel B, Bartnitzke S, Dal Cin P, van de Ven W, van den Berghe, Bullerdiek J (1991) Leiomyoma cells with 12q15 aberrations can be transformed in vitro and show a relatively stable karyotype during precrisis period. Cancer Genet Cytogenet 54:223-228

Stewart N, Bacchetti S (1991) Expression of SV40 large T antigen, but not small t antigen, is required for the induction of chromosomal aberrations in transformed human cells. Virology 180:49–57

Todaro GJ, Green H, Swift MR (1966) Susceptibility of human diploid fibroblast strains to transformation by SV40 virus. Science 153:1252–1254

Wolman SL, Hirschhorn K, Todaro GJ (1964) Chromosomal changes in SV40-infected human fibroblast cultures. Cytogenetics 3:45–61

Woodworth CD, Kreider JW, Mengel L, Miller T, Meng Y, Isom HC (1988) Tumorigenicity of simian virus 40 hepatocyte cell lines: effect of in vitro and in vivo passage on expression of liver-specific genes and oncogenes. Mol Cell Biol 8:4492–4499

Analysis of DNA from a Specific Chromosome Region

N. Ponelies[1,2], J. Scheef[1], M. Köllner[1], N. Endlich[1], and K.O. Greulich[1]

1 Introduction

Finding the locus of a gene when its mRNA or gene product is known is possible but tedious. On the other hand, isolating a gene from the cytogenetically determined chromosomal locus of a disease (positional cloning or, formerly, reverse genetics; Collins 1992) has so far been successful only in very few cases. Although the Human Genome Mapping Catalogue HGM 11 lists almost 2000 cloned genes, knowledge of the architecture of the human genome is still sporadic and statistical. This is illustrated by the fact that the haploid DNA molecule of a human cell is in total almost 1 m long, and that the total DNA in the human body if aligned one after the other is sufficient to cover the distance between earth and sun approximately 80 times. Even a small chromosome band contains a few megabases of DNA with an unpacked length in the order of millimeters, while the longest piece of DNA (*C. elegans*) which has been completely sequenced, comprises only about 122 kb (Sulston et al. 1992). Thus it becomes clear that we are far from the goal of systematically analyzing the human genome. It may become possible in the not too distant future to systematically analyze a selected chromosome band or a chromosome segment containing very few bands. For this purpose techniques must be provided to isolate DNA selectively from such a chromosome segment. Especially in human genetics and medical diagnosis new methods must be developed. Techniques using somatic cell hybrids (Wasmuth et al. 1986; Gusella et al. 1980), radiation-induced hybrids (Cox et al. 1989) or flow sorted chromosomes (Fuscoe et al. 1986; Youngman et al. 1989) have enabled the construction of chromosome-specific DNA libraries. The importance of such defined DNA probes is clearly recognizable, for example, in human genetic disease research (van Ommen et al. 1986; Nussbaum et al. 1987; Davis et al. 1990).

---

[1] Phys.-Chem. Institut, Im Neuenheimer Feld 253, Universität Heidelberg, 69121 Heidelberg, Germany

[2] Molekulare Genetik, Im Neuenheimer Feld 236, Universität Heidelberg, 69121 Heidelberg, Germany

Hybrid cells containing the segment of interest may be one source for suitable DNA. Such hybrids are, however, not available for every region of the human genome and, when prepared as radiation hybrids, are quite often unstable.

A second approach for preparing chromosome segment specific DNA probes is chromosome microtechnology, i.e., microdissection with microamplification or microcloning of the dissected material. Mechanical microdissection is conceptually straightforward but in practice requires a skilled experimenter. On the other hand, laser microdissection (Monajembashi et al. 1986; Djabali et al. 1991) requires equipment unfamiliar to the biologist, a UV laser microbeam (Wiegand et al. 1987; Weber and Greulich 1992) or a microbeam trap (Greulich 1992a; Greulich and Weber 1992). Once the apparatus is set up, however, laser microdissection is easy to learn and can therefore be performed by a person interested in the biological problem itself rather than an expert in the microdissection technique.

One drawback of the microdissection approach is the fact that only a few DNA molecules are available as starting material, i.e., very efficient amplification procedures must be employed. In the past 10 years several techniques have been developed for this purpose. Approaches based on polymerase chain reaction (PCR) techniques (Lüdecke et al. 1989; Kao and Yu 1991) are simple and easy to handle but entail some statistical hazards (Greulich 1992b; see also below). Classical cloning for example into λ phage is less problematic in this respect but requires more sophisticated experimental techniques.

In addition to the molecular biological techniques for obtaining access to DNA from a specific chromosome segment, a microscopic approach is emerging which allows direct observation of the dynamics of an individual DNA molecule. While this technique is still in a state of basic research, it can already be envisaged that direct observation of DNA has the potential to become an interesting technique for human genome research in the near future.

2 The Laser Equipment: Laser Microdissection of Chromosomes

The laser equipment consists of a Zeiss Axiovert microscope in which the fluorescence lamp is complemented by a pulsed UV laser used for ultrafine cutting and by a continuous infrared laser used as ultrafine tweezers or optical trap. A simplified representation of the microbeam trap is given in Fig. 1. For laser microdissection the pulsed UV laser with a wavelength of 337.5 nm and a pulse duration of 5 ns is used. Such a laser provides high power densities which can be used in a similar way as lasers in industrial

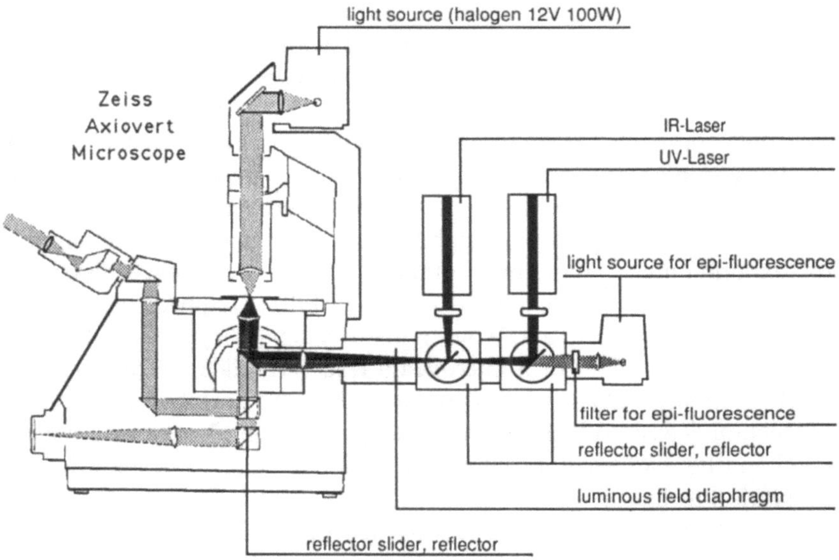

Fig. 1. The microbeam trap. A nitrogen laser providing high power ultraviolet pulses (LSI, Cambridge, MA, USA, or Laser Technik Berlin, Germany) and a continuous infrared laser (Adlas, Lübeck, Germany) are coupled via the fluorescence illumination path into an Axiovert Microscope (Zeiss, Oberkochen, Germany). The ultraviolet nitrogen laser can be used for microdissection of chromosomes; the infrared neodymium-YAG laser is used as optical tweezers. A 100 × Ultrafluar objective is used to focus the lasers onto the object plane

cutting processes, albeit with much better precision. The accuracy is limited by the wavelength of the light used. With UV light, chromosomes can be dissected with a precision of 15–30 Mb. Figure 2 shows an electron micrograph of a human chromosome cut with a nitrogen laser.

The target region cut out of metaphase spreads with the laser beam is, in most present applications, collected with a siliconized glass needle. If suspension chromosomes can be used, it is also possible to collect the microdissected segment with the infrared optical trap (Fig. 3). The working principle of the latter is similar to dielectrophoresis, the technique that preceding electrofusion of cells for collecting them. It uses the fact that dielectric objects such as chromosome segments migrate into the direction of maximum electric field, and that the focus of a laser beam always represents such a field maximum. Thus, chromosome segments can be held and moved by such an infrared laser without significantly damaging the DNA contained in the chromosome.

Recently, an alternative approach for obtaining access to isolated chromosome segments has been suggested by Djabali et al. (1991) and Hadano et al. (1991). There, whole metaphases with the exception of the target segment are ablated by the laser and the remaining segment is rinsed

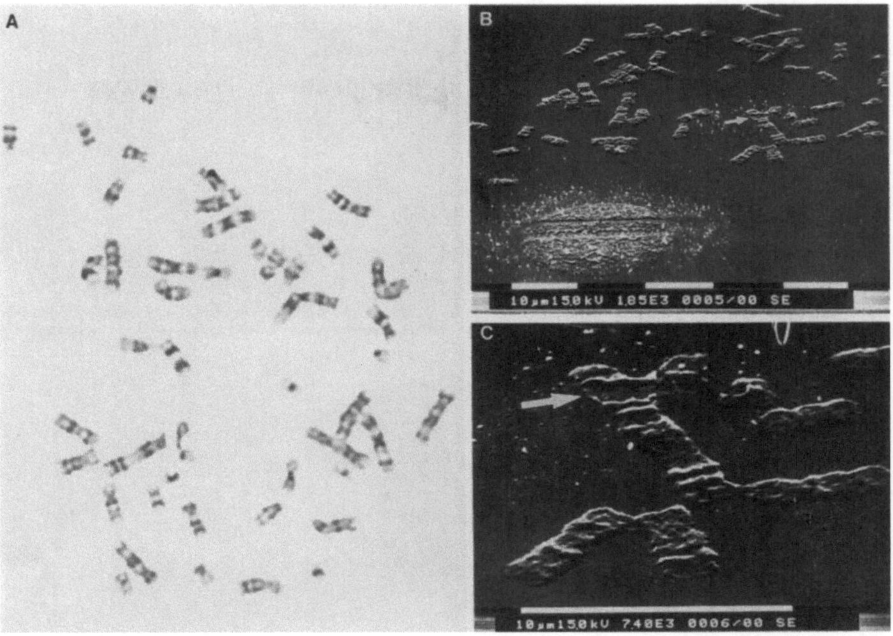

Fig. 2. **A** G-banded metaphase spread of a t(1;7)(1qter-1p36::7q11-7pter) translocation patient with a laser cut (*arrow*) on the end of 1q. The cell line was kindly provided by K. Miller (Hanover, Germany). **B,C** Scanning electron micrographs (courtesy of N. Paweletz, German Cancer Research Center, Heidelberg, Germany) of the same laser microdissected (*arrows*) chromosome shown in **A**. The observation angle is 60°. *Bar* (**B**, **C**), 10 µm

from the cover slip. The entire process also does not require micromanipulation.

3 Cloning Strategies for Selected Chromosomal Regions

For amplification of the chromosome segments after microdissection a choice of techniques is available. Vector-dependent techniques are usually referred to as microcloning, and techniques based solely on molecular amplification such as PCR and referred to as microamplification (see also Greulich 1992b). For both cases the main problem is to work with a few molecules as starting material, i.e., the substrate concentration for all enzymatic steps is several orders of magnitude smaller than in the corresponding standard procedures. For example, standard cloning procedures start with 0.1–1 µg DNA while the DNA contained in 100-chromosome segments is approximately 3 pg. Similarly, PCR amplification with single unique or *Alu* primers start with 10 ng as compared to 30 fg for the DNA of a single chromosome segment (see below, and Tables 1, 2).

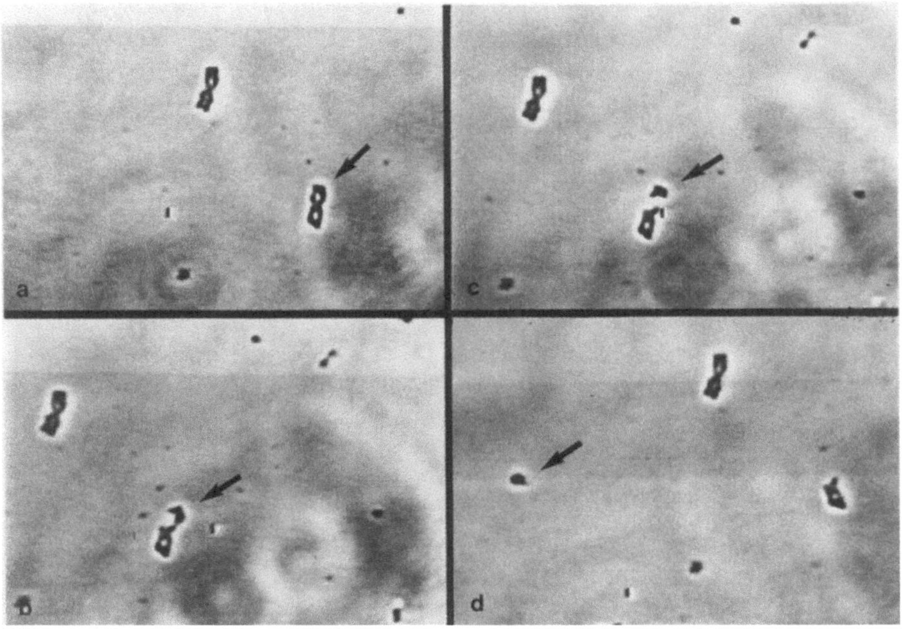

Fig. 3a–d. Use of the microbeam trap for microdissection and subsequent pooling of chromosome segments. **a** Chromosomes before laser treatment. **b** The tip of the chromosome on the right is partially cut off by a single nitrogen laser pulse. **c** Complete separation of the tip after a second pulse. **d** The tip is separated from the chromosome by fixing it in the optical trap and by moving the object stage to the right

3.1 Microcloning into λ Phage

Microcloning after microdissection with a glass needle was first performed by Scalenghe et al. (1981). This technique allows direct isolation of the chromosomal region of interest without limitation to a specific chromosome. The approach consists mainly of a special micromanipulation method which allows working in nanoliter volumes. The dissected chromosome segments are isolated and extracted in nanoliter droplets. After protein and phenol extraction and restriction of fragments, the DNA is ligated into phage vectors. All steps are optimized to obtain sufficient ligation efficiency or a high number of recombinants. The experiments are carried out with specific micromanipulation instruments under a phase contrast microscope. Microcloning after mechanical microdissection has been applied to diptera (Pirrotta et al. 1983; Delaney et al. 1991), mouse (Röhme et al. 1984; Nisswander et al. 1991), and human (Bates et al. 1986; Kaiser et al. 1987) genomes. The technique has a wide area of applications, for example, for the generation of probes in genetic diseases (Bautsch et al. 1988) and cancer research (Leister et al. 1990), regional mapping of specific chromosomes, organization and function of morphological features of chromosomes (telomeres/centromeres), and cytogenetic studies (banding pattern/ chromosome painting; Lengauer et al. 1991).

Despite the fact that the development of new DNA markers by the microcloning approach is less time consuming than preparation of chromosome segment specific DNA probes via cell hybrids or using sorted chromosomes (see Sect. 1), some critical points must be observed (see below), and the original techniques must be adapted to new demands (Table 1). For alternatives to classical microcloning in general, two types of strategies have been employed. One is to modify classical microcloning, and the other uses PCR technologies.

3.2 Modification of the Classical Microcloning Approaches

Conventional microcloning to generate chromosome segment specific probe libraries can be facilitated by modification of the microdissection step. The microcloning approach requires handling with micromanipulation instruments. Thus it is more time consuming than PCR methods, and a highly skilled experimenter is needed. To provide an easier way to handle the method, we replaced the mechanical dissection of chromosomes by laser dissection (Monajembashi et al. 1986; Ponelies et al. 1989). Laser microdissection has been complemented by optical trapping to dissect and pool chromosome pieces without mechanical micromanipulators; however, the latter at present works only after special preparation conditions.

If one uses spread metaphase chromosomes as starting material for probe generation, the localization of the desired region must be managed. For microcloning the use of banded chromosomes is not yet practical since a time-dependent degradation damages the chromosomal material.

3.3 Microamplification Based on PCR Techniques

PCR-based approaches are sensitive, and one can start with extremely low amounts of DNA. The yields of the PCR products are high, and the experiments are straightforward. Lüdecke et al. (1989) have cloned *Rsa*I restriction digests into a plasmid derivative of pUC and have used known sequences in the plasmid as PCR primers. Also, PCR experiments using adapter ligation, described by Saunders et al. (1989) and Kao and Yu (1991), generate large libraries with a small number of chromosome segments. In adapter PCR, specifically designed sequences are ligated to given restriction sites by T4 ligase and are used as PCR primers. Another modification of the PCR technique is the single unique primer (SUP) method, developed by Hadano et al. (1991). SUP PCR is a three-step DNA amplification which uses a specific primer (5′ TAG ATC TGA TAT CTG

AAT TCC C 3′) different from *Alu* primers (see Sect. 3.4), able to bind randomly to human DNA and carrying restriction sites for the restriction enzymes *Eco*RI, *Eco*RV and *Bgl*II. The combination of three PCR protocols, differing mainly in the stringency of annealing reactions, the use of polyethylene glycol, and the special primer construct, makes it possible to obtain region-specific probes from only a few tens of femtograms.

On the other hand, all PCR approaches because of their high sensitivity are critical to contamination of the amplification reaction. Experiments must be performed under similar clean room conditions as cell culture experiments (Heinrich 1991). In practice, PCR products obtained after microamplification with PCR have an average size of 150–400 bp. For some applications such small probe sizes are not suitable. Also, high recombination probability and a great redundancy (see below) make it difficult to perform a systematic analysis on a defined chromosomal region. The error rate of the PCR (Saiki et al. 1988; Maruyama 1990) may prevent direct analysis of the PCR product, for example, in mutagenesis studies. The latter problem may be overcome by the use of polymerases with small error rate, as these have recently become available commercially.

3.4 Alu PCR

The PCR techniques described in Sect. 3.3 can in principle provide DNA material from any part of a genome. If in the case of the human genome only a few probes from a specific chromosome segment are needed, *Alu* PCR is a quick and simple technique. *Alu* sequences are frequently occurring repetitive sequences in the human genome ($3–9 \times 10^5$ per haploid human genome (Korenberg and Rykowski 1988)), which have an average distance of about 4–5 kb (Britten et al. 1988). The distance may vary since *Alu* sequences appear to be enriched in R bands and deficient in G/Q bands. The *Alu* family is the major family of the short interspersed repeated sequences and is approximately 3%–9% of the whole human genome and corresponds to the R bands. The *Alu* consensus sequence is about 300 bp long and has two highly conserved regions. These sequences can be used as convenient primers for subsequent amplification by PCR and cloning (Nelson et al. 1989).

Several applications and publications are listed in Table 1. Various attempts have been made to amplify DNA by *Alu* PCR from a unique chromosome. One possibility is the use of somatic cell hybrids with a single human chromosome on rodent background, as mentioned above (Nelson et al. 1989). Another is the use of flow sorted chromosomes (Cotter et al. 1991). Inspection of Table 1 shows that the DNA amounts used as starting material in *Alu* PCR are at least three orders of magnitude higher than the amount of DNA contained in a single chromosome segment (experiments listed above the bold line in Table 1). Therefore, the sensitivity of *Alu* PCR

Table 1. Conditions of various *Alu* PCR approaches

References	Total mass of DNA	Primer	Mg^{2+}	Duration/ temperature	Number of cycles	Taq polymerase
Nelson et al. (1989)	1 μg	1 μ*M* 0.01 μ*M*	1.5 m*M*	1 min/94°C 45 s/55°C 5 min/68°C	35	2.5 U
Cotter et al. (1990)	100 ng	2 μ*M*	1.5 m*M*	45 s/94°C 30 s/57°C 2 min/72°C	30–40	1.0 U
Cotter et al. (1990)	500 seg. ~18 pg	2 μ*M*	1.5 m*M*	45 s/94°C 30 s/57°C 2 min/72°C	30–40	1.0 U
Djabali et al. (1991)	1 seg. ~30 fg	2 μ*M* 2.8 μ*M* (2)	1.5 m*M* 2.1 m*M* (2)	2 min/94°C 1 min/62°C 3 min/72°C	2 × 30 (1[b])	2.5 U 2.5 U (2)
Scheef[a] et al. (in preparation)	1 seg. ~30 fg	1.8 μ*M*	12 m*M*	1 min/94°C 1.5 min/52°C 2.5 min/72°C	40	2.5 U

dNTP, 200–300 μ*M*; volume, usually 100 μl; buffer, 50 m*M* KCl, 10 m*M* Tris-HCl (pH 8.0–9.0), 0.01% gelatine.
[a] 1 m*M*.
[b] 50 μl.

must be increased to be useful for the amplification of single microdissected chromosome segments (experiments listed below the bold line in Table 1).

For standard conditions the lower limit is at about 100 pg–1 ng. A variation in the concentrations of $MgCl_2$, primer, and nucleotides shows that there is a need for higher concentrations of these compounds. In our hands, the best results were obtained with 12 mM $MgCl_2$, 1.8 μ*M* primer, and 1 m*M* nucleotides. Our primer (5′ TGC ACT CCA GCC TGG G 3′; J. Scheef et al., in preparation), which seems to be the most efficient of those that we have used, is a 16-bp oligonucleotide corresponding to the end of the *Alu* consensus sequence. The higher sensitivity resulting from changed conditions is concomitant with an increased risk of contamination. There are different methods to remove or to avoid contaminations. A very simple method is the use of filter tips (Stratagene, Heidelberg, Germany) for pipetting the different solutions. The use of these tips reduces the risk of carry-over from DNA by aerosols.

Another method is UV irradiation (λ = 254 nm) of the reaction vessels and the buffer, but due to the high absorption (transmission ~6%) of the tube material this is not very efficient. Most efficient is the treatment of amplification buffer as well as nucleotides and primer with DNase I (Boehringer Mannheim, Germany; 5 U, 45 min, 37°C). This treatment allows reliable amplification of single chromosomes and chromosome segments. This can be seen in Fig. 4, which shows the results of an *Alu* PCR

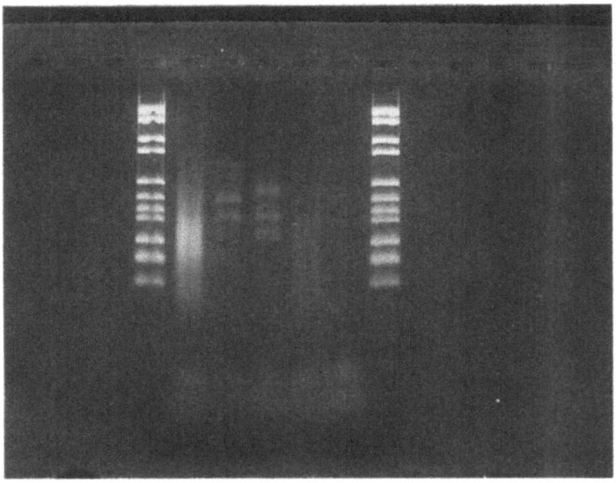

Fig. 4. The results of microamplification by *Alu* PCR. *A*, *G*, Molecular weight markers (standard marker VI; Boehringer Mannheim, Germany); 2176, 1766, 1230, 653, 517, 453, 394, 298, 234/220, and 154 bp; *B*, result of the amplification of a whole metaphase; *C*, *D*, several bands obtained after laser microdicassection and *Alu* PCR of 5 and 1 chromosome segments of chromosome 7q; *E*, positive control, 100 fg human placenta DNA (Sigma, Deisenhofen, Germany); *F*, negative control

with a whole metaphase and with five and one laser dissected chromosome segments as starting material, respectively. One sees that there are different bands on the agarose gel, and the southern hybridization experiment shows that we have in fact amplified human DNA.

While *Alu* PCR is simple to perform and highly efficient, DNA probes can be prepared only from DNA sequences intercalated between two corresponding *Alu* primer binding sites. To obtain access to any DNA sequence, alternative cloning or amplification approaches must be employed.

3.5 Microcloning Versus Microamplification

Some experiments using microcloning or microamplification are listed in Table 2. Quantitative and qualitative differences of the resulting DNA probes for both techniques can be summarized as follows. In microcloning experiments the clone number is below 1000, the average insert size is 2–3 kb, and the redundancy is small. Single copy probes obtained from such minilibraries are long enough for direct analytical studies such as hybridization experiments. Although it is not possible to cover a specific chromosomal segment completely with probes from one cloning experiment (average coverage 50–100 kb/probe), such probes may become a convenient

Table 2. DNA probe generation techniques (chromosome segment specific/human) and their properties

Method	Microdissection	Amount of DNA/ no. of segments, banding	Probe generation	No. of recombinants	Insert size (bp)	Insert size	Repetition (%)	Redundancy (%)	Reference
Microcloning	Mech. laser (dye, N_2)	1–4 pg/75–100, unstained	Microcloning procedure, vector system λ insertion	100–500	500–8000	2000–3000	20–40	<5	Weber et al. (1989) Eckelt et al. (1989)
PCR mod. microcloning	Mech.	300–500 fg/ 20–40, banded	Mod. microcloning, vector system plasmid, Sma vector, PCR(standard)/seq.-rev. seq.primer, cloning into plasmid o. λ	5000–8 × 10^4	52–400	150–180	20–30	–	MacKinnon et al. (1990) Hampton et al. (1991)
Adaptor PCR mod. microcloning	Mech. laser (N_2)	100–300 fg/ 10–30, unstained	Mod. microcloning, adaptor-ligation/no vector system, PCR (standard)/ adaptor primer, cloning into plasmid	1000–7 × 10^5	300–1300	420	40	–	Saunders et al. (1989) Kao and Yu (1991)
SUP PCR	Laser (argon ion)	10–20 fg/2, banded	No microcloning SUP PCR/three-step DNA amplification, primer BVE22cc, cloning into λ	6 × 10^6	150–800	300	55	99.985	Hadano et al. (1991)
Alu PCR	Laser	10 fg/1	No microcloning PCR (standard)/*alu* primer	–	100–800	–	–	–	Djabali et al. (1991)

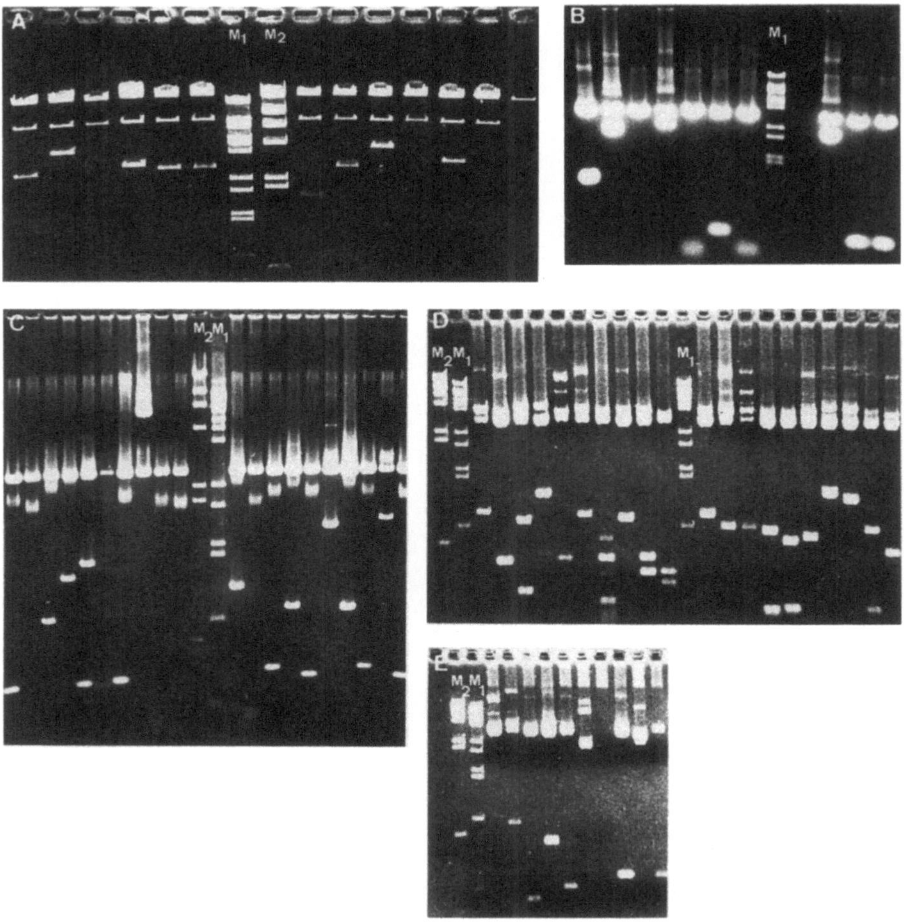

Fig. 5 A–E. Insert size distribution of different cloning approaches for generating chromosome segment specific probes. **A** Modified microcloning: probes obtained from cystic fibrosis region in λ phage NM641/laser dissection. **B** Modified microcloning: recombinants obtained with λ-Zap vector (Phagemid/Stratagene, Heidelberg, Germany) after plasmid excision. **C** Modified microcloning: using the plasmid bluescript (Stratagene, Heidelberg, Germany) as vector for microcloning. **D** Generation of chromosome 1q probes with adapter PCR; size distribution after cloning PCR-product into bluescript. **E** Modified microamplification of chromosome 1qter with "plasmid" PCR; after microcloning with bluescript the ligation product was amplified with sequencing and reverse sequencing primer, size distribution after cloning PCR product into bluescript; M1-λ-Marker/*Bst*EII, M2-λ-Marker/*Hin*dIII

source to arrange for example a number of YAC clones for mapping a distinct chromosomal region. Figure 5 serves as a comparative example for different cloning approaches and demonstrates the variable insert size distribution. It shows that using the insertion λ phage vector system one can obtain very large recombinants, but also that the average fragment size depends on the vector system used. The distinction to PCR mediated approaches is also obvious. Here the size distribution ranges between less than 100 and 1300 bp while the method of adapter PCR yields a good average size of more than 400 bp.

4 Statistical Hazards in Characterization of PCR-Based Libraries

Microcloning and particularly PCR-based techniques, useful as they are, entail several hazards due to statistical problems. In many cases it is impossible completely to characterize a microlibrary, which may contain up to 700 000 clones (Kao and Yu 1991; Table 2). Therefore, small statistical samples are characterized by one of the two following approaches: (a) To obtain information on the redundancy of the library, selected clones are hybridized against filters containing a replica of the whole library. The number of hybridization signals gives the number of identical counterparts of the test clone in the library, i.e., its redundancy (Hadano et al. 1991). (b) A statistical sample of 50–80 clones is tested for repetitiveness, clone size, and sometimes partial sequence. From the composition of this statistical sample a linear extrapolation is taken to characterize the whole library (Kao and Yu 1991).

The following considerations show that such extrapolations may lead to erroneous conclusions. Because of the importance of this aspect we illuminate, below, some mathematical details of the underlying statistics. For a less mathematical qualitative consideration see Greulich (1992b).

4.1 Statistics of Filter Hybridization

The aim is the estimation of the number of different clones on a filter by hybridization when the total number of clones on the filter is known. The total number of clones N is

$$N = c_1 + c_2 + \dots \qquad (1)$$

where c_i is the number of copies of clone i. If one arranges all clones with the same number of copies c_i together and finds S_i clones with c_i copies, one can rewrite N as

$$N = S_1c_1 + S_2c_2 + \dots \qquad (2)$$

If now the total number of different clones $S = S_1 + S_2 + \dots$, in which one is interested, is extracted, one finds:

$$N = S(S_1/Sc_1 + S_2/Sc_2 + \dots) \qquad (3)$$

But S_i/S is the fraction of clones with c_i copies. Thus the sum is merely the average number of copies on the entire filter or

$$N = S\langle c \rangle \tag{4}$$

where $\langle\rangle$ denotes the average value. A good estimate for S after hybridization and counting the number of hybridization signals in the sample would therefore be

$$S^* = N/\langle c \rangle_{\text{sample}} \tag{5}$$

where * marks the estimate of S, and the average is taken over the hybridized sample.

A trimodal distribution of c_i may serve as an example. A library may consist of 100 clones with 900 copies each, 100 clones with 90 copies each, and 1000 single copy clones. N is then 100 000, $S = 1200$, $c_1 = 900$, $c_2 = 90$, and $c_3 = 1$. If only one clone is hybridized against the filter, the probability of obtaining only one hybridization signal is very high despite the fact that the total library is highly redundant. The probability of such a result is 1000/1200 or 83%. In this case one would guess 100 000 for S. In other words, in 83% of the experiments one would estimate the value of S two orders of magnitude too large!

In Fig. 6 the probability of a certain guess S^* for S is plotted for five hybridizations. One clearly sees that even with more than one hybridization the guess S^* for S is usually far from the true value 1200. In this particular example the probability of obtaining a good estimate is obviously very low.

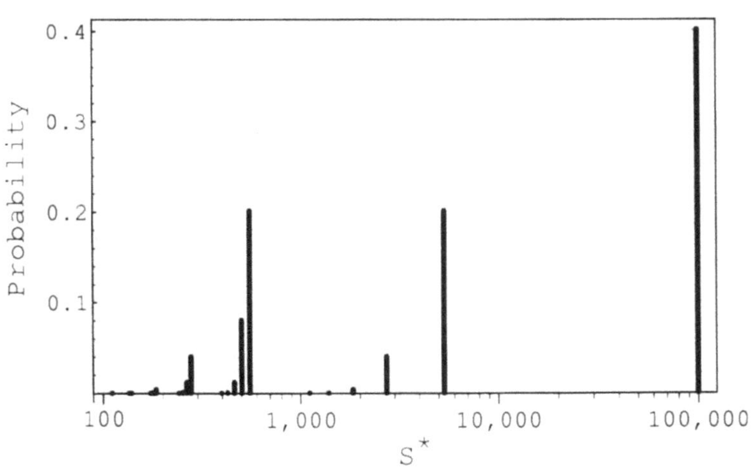

Fig. 6. Statistics: the probability of a certain estimate S^* for S, the number of different clones on a filter, after five hybridizations is plotted in a histogram for the example of a trimodal distribution of the abundance of the clones: 100 clones with 900 copies, 100 clones with 90 copies and 1000 clones with only 1 copy. The correct resut for S would be 1200. The probability of obtaining this result is very small

In conclusion, one must use a large sample to obtain a reliable estimate of the number of different clones. Even for 100 hybridizations the half-width of the histogram for the above example extends from $S^* = 900$ to 2000, so the relative error is still near 100% (!).

4.2 Problems with Small Samples for Characterization of Large Libraries

A numerical example can illustrate the problem with small samples. If a sample of 50 clones out of a library containing 20000 clones is used for its characterization, and if the result is that 40 clones are different, then by linear extrapolation one may be tempted to conclude that 16000 clones in the library are different. We will prove this procedure to be invalid and will show that by a more detailed statistical analysis significantly different estimates of the number of different clones in the library may be obtained. The problem encountered here is a sampling problem. A complete analysis of it is provided by Watterson (1974), with many further references cited. Here the main features of the problem and its solution are outlined briefly.

Outline of the Problem. If one takes a sample with only a part of the total clone number, and if one wants to estimate the total number of different clones in the library, one would best give an estimate including confidence intervals. In many cases this is rather difficult. It makes a difference whether the library is make up of only a few different clones, each of which is represented by many copies, or if the library is composed of many different clones of only very few copies each. While in the first case even a small sample is sufficient to estimate the number of different clones in the library, in the latter case one would need a large sample to obtain a fairly accurate estimate. Further, if the library is composed of some types of clones in high abundance, and many other types in small proportion, it is difficult to detect the number of clones with small proportion or below a certain abundance, i.e., some of the most important clones. In general, a good approach would be to decide sequentially: take a small sample, analyze it, and decide whether further sampling is necessary, and so on. For an exact treatment of the general sampling and estimating problem one would at first have to establish a model for the distribution of the clones and their proportions in the library, then determine the distribution of the clones in the sample, and from this knowledge develop an estimator for the number of different clones and their proportions in the library.

Example. The analysis is restricted to the case in which an equal number of copies are present for all the different clones. Let N be the library size, S the number of different clones in the library, n the number of clones tested, i.e., the sample size, and s the number of different clones in the sample. For this case $\langle s \rangle$, the mean value of s, can be given as a function of n and the desired parameter S (Watterson 1974):

$$\langle s \rangle = S\,[1 - (1 - 1/S)^n\,] = f_n\,(S) \tag{6}$$

For large S and n expressions such as $(1 - 1/S)^n$ are approximately $e^{-n/S}$; thus Eq. 6 is simplified to:

$$\langle s \rangle \approx S\,(1 - e^{-n/S}) \tag{6a}$$

If n/S is large, $E(s) \approx S$, and if n/S is very small, $E(s) \approx n$. This can be seen in Fig. 7. In a crude manner of estimating S one might guess that $s_{\text{measured}} \approx \langle s \rangle$ and solve $s_{\text{measured}} \approx f_n(S)$ for S giving:

$$S^* \approx f_n^{-1}(s_{\text{measured}}) \tag{7}$$

where, again, * marks the estimate of S. The example mentioned at the beginning may serve as illustration: the parameters were $n = 50$ and $s = 40$. A rough estimate for s would therefore be $S^* \approx 100$, which is more than two orders of magnitude less than the intuitive estimate. These results have been supported by simulation which should be used for this problem as long as detailed analytical solutions are lacking. The most important feature of the solution is the fact that the number of different clones in the library is not linearly related to the number of different clones in the sample. In no case can S be calculated in a linear manner by $N \cdot s/n$. The library size N is actually completely irrelevant, as long as it is large; it does not appear in any of the formulas.

5 Observation of Single DNA Molecules in the Light Microscope

So far we have reported on the access to DNA from a specific chromosome region via microcloning or microamplification. There is, however, a technique that is emerging which allows a much more direct access to single DNA molecules: observation by light microscopy. At present, only model DNA, for example from λ phage, T4 phage, calf thymus, or cosmid has been investigated, but it is obvious that laser microdissection will pave the way to human DNA segments in the near future.

Morikawa and Yanagida (1981) and Matsumoto et al. (1981) have reported on such a fluorescence technique by which not only highly condensed metaphase chromosomes can be visualized in the light microscope but also individual, naked DNA molecules down to a size of 10 kb. Thus, it is possible to study the dynamic behavior of single DNA molecules (Yanagida et al. 1982; Houseal et al. 1989) stained with a fluorescence dye directly in solution, although the diameter of DNA molecules is below the resolution limit of the light microscope. This surprising effect works by the same principle that enables us to see the stars in the night; we can see the emitted light of each star in the dark sky, but we cannot spatially resolve structural details. With this in situ observation technique Yanagida et al. (1982) and Houseal et al. (1989) observed the

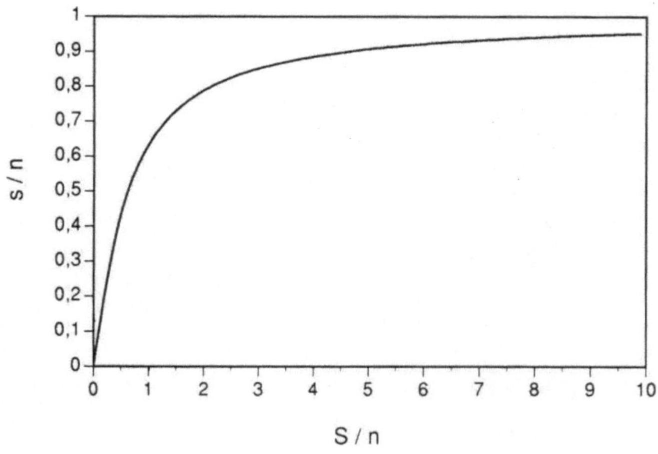

Fig. 7. The influence of *S*, the number of different clones in the library, on *s*, the number of different clones in the sample (in units of *n*, the sample size), for the case where an equal number of copies is present for all the different clones. The library size is completely irrelevant as long as it is large, and *S* and *s* are not linearly related

reversible structural transitions of rodlike DNA molecules into globular or ellipsoidal shapes on a time scale of seconds. With respect to biological functions (Williamson and Fennel 1975; Hirano and Mitchison 1991) we have adopted and modified this technique to study the dynamics of single DNA molecules.

Our preliminary results are summarized as follows. Immediately after preparation the DNA molecules are elongated and appear as long filaments, probably caused by external forces applied on the DNA containing solution during pipetting. However, without external forces, they tend to condense into a compact structure. In the presence of Ficoll the time constants for the condensation in aqueous solution were minutes rather than seconds without Ficoll. A typical condensation process of λ phage concatamers stained with 4′,6-diamidinophenylindole (DAPI) is shown in Fig. 8.

Elongated DNA molecules frequently adhere to the microscope slide at one or more points. Then, during the condensation process intramolecular tensions increase and finally disrupt the molecule, whereas each fragment continues the condensation process. The globular shape reveals fluctuations between spherical, ellipsoidal, and irregular structures (Fig. 9).

Using the single molecule technique, we have observed in situ the disintegration of individual DNA molecules as a result of induction of nicks by UV irradiation (λ = 254 nm). It may therefore become possible, especially in view of chromosomal abnormalities in human cancer, such as translocations, deletions, and insertions (Mitelman et al. 1991; Dal Cin and Sandberg 1989; Sandberg 1991; Seizinger et al. 1991), to investigate whether chromosome damage may be caused by a physical defect in the DNA

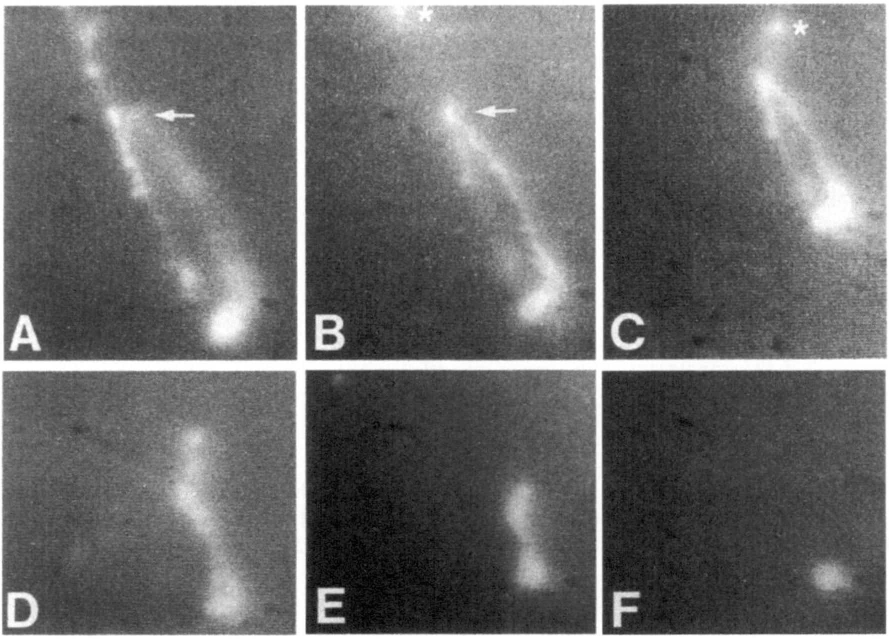

Fig. 8 A–F. An individual, DAPI-stained (final concentration of 1 μg/ml), λ phage concatamer in a Ficoll-containing solution. One end of the DNA (*arrow*, **A**) pulled near the extended part of the molecule and formed a loop (*arrow*, **B**). The upper part of the DNA molecule (*asterisks*, **B**, **C**) began with the condensation process. During the condensation the filament assumes an increasingly rodlike structure (**D**, **E**) and finally collapses into a globular molecule (**F**). ×1600

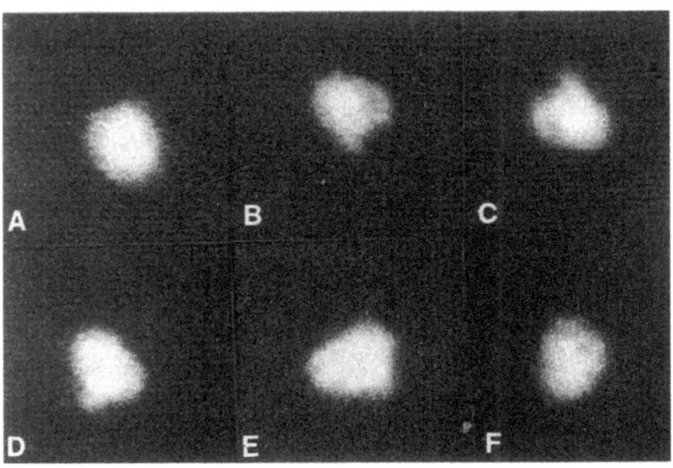

Fig. 9 A–E. Internal structural fluctuations of a globular molecules. The single molecule randomly fluctuated between spherical (**A**), ellipsoidal (**B**), and irregular (**B–E**) structures on a time scale of 14–25 s. ×2300

molecule. It may also be interesting to study directly the interaction of such DNA molecules with histone molecules in order to learn more about DNA protein interaction on the basic stage of chromosome formation.

To study such an interaction between various histone proteins and DNA several indirect methods are commonly used, for example, X-ray diffraction and circular dichroism, but so far electron microscopy is the only imaging technique (Felsenfeld 1978; Thoma et al. 1979; Thoma and Koller 1981). However, these methods are difficult to use when the dynamics of the DNA molecule is to be investigated. The fluorescence technique allows one to study the reaction between individual DNA molecules and histones or other polycations directly in solution without denaturing preparation. The influence of the polycation poly-L-lysine and the histones on the structure of λ phage concatamers in solution is shown in Fig. 10. After the addition of poly-L-lysine (final concentration of 0.1 μg/ml) the DNA (100 μg/ml) either condensed into small compact particles or formed long interwoven filaments, depending on the preparation procedure. The histones H_1, H_2, and H_3 (final concentration 0.13 mg/ml) also induced condensation. In contrast to poly-L-lysine, bloated structures and also long interwoven filaments are observed (Fig. 10). In both cases the DNA was frequently fixed on the surface of the microscope slide. It may be interesting to read the histone H_1 to H_1-depleted chromatin to reorganize the chromosome structure, as has been done in an electron microscopy study (Thoma and Koller 1981).

6 Concluding Remarks

There is now a choice of techniques to obtain access to DNA from a specific chromosome segment. By touching on some aspects of the different cloning approaches described here, we have tried to make apparent two important points. First, at the present time there is no universal technique available to generate specific DNA probes. Second, all of the discussed methods work with reasonable efficiency, but for each application a careful choice of the specific approach is necessary. Our own efforts have concentrated on simplifying the experimental technique, introducing alternative vector systems and performing comparative studies of microcloning with PCR microamplification.

Microscopic techniques for studying the dynamics and physical properties of DNA on a single-molecule basis may soon complement the molecular biological approaches. What remaïns for the future is the development of single molecule hybridization techniques and laser microdissection and handling of these individual molecules with the microbeam trap.

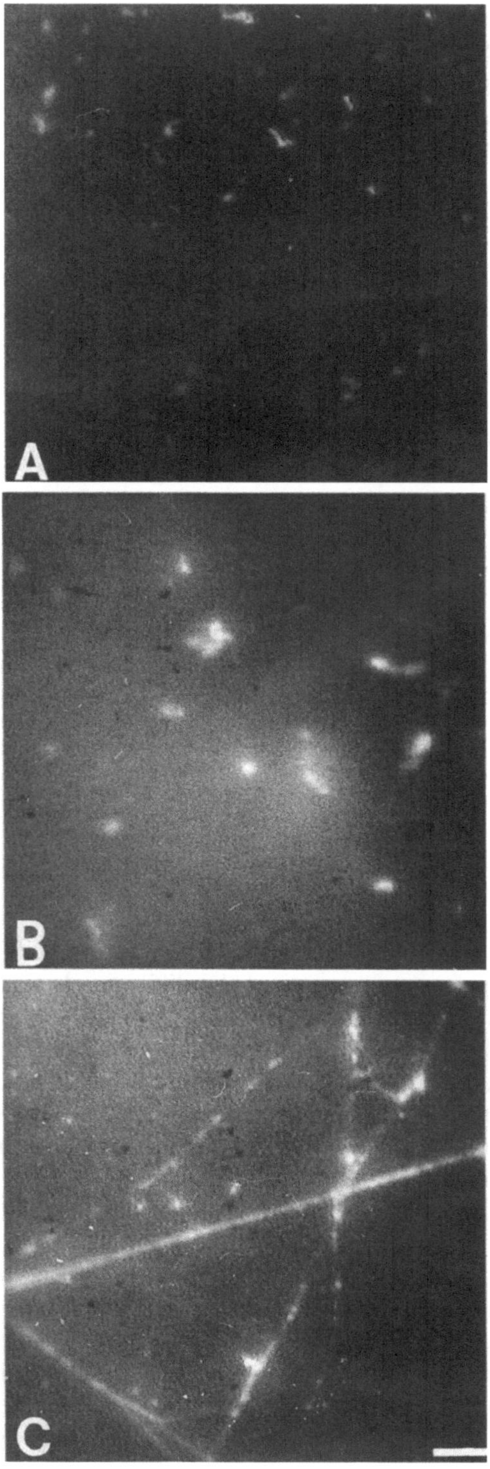

Fig. 10. A Poly-L-lysine 50000 (1 μl; Serva, Heidelberg, Germany; 5 mg/ml H_2O) was added to 5 μl DAPI-stained λ concatamer solution (Biometra, Göttingen, Germany; 0.1 mg/ml DNA in 2*SSC buffer which also contained 0.025% orange G and 3% Ficoll) and stored at room temperature for 20 min; 5 μl of the solution was placed on a microscope glass slide and covered with a cover slip. **B** Histone solution (1 μl; lysine-rich H_1, 2 mg/ml; slightly lysine-rich H_2, 1 mg/ml; arginine-rich H_3, 1 mg/ml in H_2O; Sigma, Deisenhofen, Germany) was added to 5 μl λ concatamer solution. After 20 min the mixture was placed on a microscope slide and covered with a cover slip. *Scale bar*, 10 mm. **C** The λ phage solution (5 μl) was placed on a microscope slide; 1 μl histone H_1 (Sigma, Deisenhofen, Germany; 2 mg/ml H_2O) was added and furnished directly with a cover slip

References

Bates GP, Wainwright BJ, Williamson R, Brown SDM (1986) Microdissection of and microcloning from the short arm of human chromosome 2. Mol Cell Biol 6:3826–3830

Bautsch B, Ponelies N, Darnedde T, Fryburg K, Grothues D, Hundrieser J, Miller K, Monajembashi S, Claaß A, Greulich KO, Maaß G, Tümmler B (1988) The nasal polyps as a tool for basic research in cystic fibrosis. Scand J Gastroenterol 143:5–8

Britten RJ, Baron WF, Stout DB, Davidson EH (1988) Sources and evolution of human Alu repeated sequences. Proc Natl Acad Sci USA 85:4770–4774

Collins FS (1992) Positional cloning: let's not call it reverse any more. Nature Genet 1:3–6

Cotter FE, Hampton GM, Nasipuri S, Bodmer WF, Young BD (1990) Rapid isolation of human chromosome-specific DNA probes from a somatic cell hybrid. Genomics 7:257–263

Cotter FE, Das S, Donek E, Carter NP, Young BD (1991) The generation of DNA-probes to chromosome 11q23 by Alu PCR on small numbers of flow-sorted 22q-derivative chromosomes. Genomics 10:473–480

Cox DR, Pritchard CA, Uglum E, Casher D, Kobori J, Myers RM (1989) Segregation of the Huntingtion disease region of human chromosome 4 in a somatic cell hybrid. Genomics 4:397–407

Dal Cin P, Sandberg AA (1989) Chromosomal aspects of human oncogenesis. Crit Rev Oncog 1:113–126

Davis LM, Senger G, Lüdecke HJ, Claussen U, Horsthemke B, Zhan SS, Metzroth B, Hohenfellner K, Zabel B, Shows TB (1990) Somatic cell hybrid and long-range physical mapping of 11p13 microdissected genomic clones. Proc Natl Acad Sci USA 87:7005–7009

Delaney SJ, Hayward DC, Barleben F, Fischbach KF, Miklos GLG (1991) Molecular cloning and analysis of small optic lobes, a structural brain gene of Drosphila melanogaster. Proc Natl Acad Sci USA 88:7214–7218

Djabali M, Nguyen C, Biunno I, Oostra BA, Mattei MG, Ikeda JE, Jordan BR (1991) Laser microdissection of the fragile X region: identification of cosmid clones and of conserved sequences in this region. Genomics 10:1053–1060

Eckelt A, Ponelies N, Bautz EKF, Miller K, Heuer T, Tümmler B, Grzeschik KH, Wolfrum J, Greulich KO (1989) Laser microdissection in the search for the molecular basis of diseases: a molecular library from a human chromosome segment. Ber Bunsenges Phys Chem 93:1446–1450

Felsenfeld G (1978) Chromatin (review article). Nature 271:115–121

Fuscoe JC, Clark LM, Van Dilla MA (1986) Construction of 15 human chromosome-specific DNA libraries from flor-purified chromosomes. Cytogenet Cell Genet 43:79–86

Greulich KO (1992a) Moving particles by light: no longer science fiction. R Microsc Soc Proc 27:3–8

Greulich KO (1992b) Chromosome microtechnology: microdissection and microcloning. Trends Biotechnol 10:48–51

Greulich KO, Weber G (1992) The light microscope on its way from an analytical to a preparative tool. J Microsc (in press)

Gusella JF, Keys C, Varsanyi-Breiner A, Kao FT, Jones C, Puck TT, Housman D (1980) Isolation and localization of DNA segments from specific human chromosomes. Proc Natl Acad Sci USA 77:2829–2833

Hadano S, Watanabe M, Yokoi H, Kogi M, Kondo I, Tsuchiya H, Kanazawa I, Wakasa K, Ikeda JA (1991) Laser microdissection and single unique primer PCR allow generation of regional chromosome DNA clones from a single human chromosome. Genomics 11:364–373

Hampton G, Leuteritz G, Lüdecke HJ, Senger G, Trautmann U, Thomas H, Solomon E, Bodmer WF, Horsthemke B, Claussen U, Ballhausen WG (1991) Characterization and

mapping of microdissected genomic clones from the adenomatous polyposis coli (APC) region. Genomics 11:247–251
Heinrich M (1991) PCR carry-over. Biotech Forum Europe 8
Hirano T, Mitchison TJ (1991) Cell cycle controll of higher-order chromatin assembly around naked DNA in vitro. J Cell Biol 115:1479–1489
Houseal TW, Bustamante C, Stump RF, Maestre MF (1989) Real-time imaging of single DNA molecules with fluorescence microscopy. Biophys J 56:507–516
Kaiser R, Weber J, Grzeschik KH, Edström JE, Driesel A, Zengerling S, Buchwald M, Tsui C, Olek K (1987) Microdissection and microcloning of the long arm of human chromosome 7. Mol Biol Rep 12:3–6
Kao FT, Yu JW (1991) Chromosome microdissection and cloning in human genome and genetic disease analysis. Proc Natl Acad Sci USA 88:1844–1848
Korenberg JR, Rykowski MC (1988) Human genom organisation: alu, Lines and the molecular structure of metaphase chromosome bands. Cell 53:391–400
Leister I, Weith A, Brüderlein S, Cziepluch C, Kangwanpong D, Schlag P, Schwab M (1990) Human colorectal cancer: high frequency of deletions at chromosome 1p35. Cancer Res 50:7232–7235
Lengauer C, Eckelt A, Weith A, Endlich N, Ponelies N, Lichter P, Greulich KO, Cremer T (1991) Painting of defined chromosomal regions by in situ suppression hybridization of libraries from laser-microdissected chromosomes. Cytogenet Cell Genet 56:27–30
Lüdecke HJ, Senger G, Claussen U, Horsthemke B (1989) Cloning defined regions of the human genome by microdissection of banded chromosomes and enzymatic amplification. Nature 338:348–350
MacKinnon RN, Hirst MC, Bell MV, Watson JEV, Claussen U, Lüdecke HJ, Senger G, Horsthemke B, Davis KE (1990) Microdissection of the fragile X region. Am J Hum Gen 47:181–187
Maruyama IN (1990) Estimation of errors in the polymerase chain reaction. PCR Technique 2:302–303
Matsumoto S, Morikawa K, Yanagida M (1981) Light microscopic structure of DNA in solution studied by the 4′,6-diamidino-2-phenylindole staining method. J Mol Biol 152:501–516
Mitelman F, Kaneko Y, Trent J (1991) Report of the committee on chromosome changes in neoplasia. Cytogenet Cell Genet 58:1053–1079
Monajembashi S, Cremer C, Cremer T, Wolfrum T, Greulich KO (1986) Microdissection of human chromosomes by a laser-microbeam. Exp Cell Res 167:262–265
Morikawa K, Yanagida M (1981) Visualization of individual DNA molecules in solution by light microscopy: DAPI staining method. J Biochem 89:693–696
Nelson DL, Ledbetter SA, Corbo L, Victoria MF, Ramirez-Solis R, Webster TD, Ledbetter DH, Caskey CT (1989) Alu polymerase chain reaction: a method for rapid isolation of human-specific sequences from complex DNA sources. Proc Natl Acad Sci 86:6686–6690
Nisswander L, Kelsey G, Schedl A, Ruppert S, Sharan SK, Holdener-Kenny B, Rinchik EM, Edström JE, Magnuson T (1991) Molecular mapping of albino deletions associated with early embryonic lethality in the mouse. Genomics 9:162–169
Nussbaum RL, Lesko JG, Lewis RA, Ledbetter SA, Ledbetter DH (1987) Isolation of anonymous DNA sequences from within a submicroscopic X chromosomal deletion in a patient with choroideremia, deafness, and mental retardation. Proc Natl Acad Sci USA 84:6521–6525
Pirrotta V, Hadfield C, Pretorius GHJ (1983) Microdissection and cloning of the white locus and the 3B1–3C2 region of the Drosophila X chromosome. EMBO J 2:927–934
Ponelies N, Bautz EKF, Monajembashi S, Wolfrum J, Greulich KO (1989) Telomeric sequences derived from laser-microdissected polytene chromosomes. Chromosoma 98:351–357
Röhme D, Fox H, Hermann B, Frischauf A, Edström JE, Mains P, Silver LM, Lehrach H (1984) Molecular clones of the mouse t-complex derived from microdissected metaphase chromosome. Cell 36:783–788

Saiki RK, Gelfand GH, Stoffel S, Scharf SJ, Higuchi R, Horn GT, Mullis KB, Ehrlich HA (1988) Primer directed amplification of DNA with a thermostable DNA polymerase. Science 239:487–491

Sandberg AA (1991) Chromosomes abnormalities in human cancer and leukemia. Mutat Res 247:231–240

Saunders RDC, Glover DM, Ashburner M, Kiamos IS, Louis C, Monastirioti M, Savakis C, Kafatos F (1989) PCR amplification of DNA microdissected from a single polytene chromosome band: a comparison with conventinal microcloning. Nucleic Acids Res 17

Scalenghe F, Turco E, Edström JE, Pirrotta V, Melli M (1981) Microdissection and cloning of DNA from a specific region of Drosophila melanogaster polytene chromosomes. Chromosoma 82:205–216

Seizinger BR, Klinger HP, Junien C, Nakamura Y, Le Beau M, Cavenee W, Emanuel B, Ponder B, Mitelman F, Louis D, Menon A, Newsham I, Decker M, Kaelbing M, Henry I, Deimling AV (1991) Report of the committee on chromosome and gene loss in human neoplasia. Cytogenet Cell Genet 58:1080–1096

Sulston J, Du Z, Thomas K, Wilson R, Hillier L, Staden R, Halloran N, Green P, Thierry-Mieg J, Qiu L, Dear S, Coulson A, Craxton M, Durbin R, Berks M, Metzstein M, Hawkins T, Ainscough R, Waterston R (1992) The C. elegans genome sequencing project: a beginning. Nature 356(6364):37–41

Thoma F, Koller T (1981) Unravelled nucleosome beads and higher order stuctures of chromatin: influence of non-histone components and histone H1. J Mol Biol 149:709–733

Thoma F, Koller T, Klug A (1979) Involvement of histone H1 in the organisation of the nucleosome and of the salt-dependent superstructures of chromatin. J Cell Biol 83:403–427

van Ommen GJB, Verkerk JMH, Hofker MH, Monaco AP, Kunkel LM, Ray P, Worton R, Wieringa B, Bakker E, Pearson PL (1986) A physical map of 4 million bp around the Duchenne Muscular Dystrophy gene on the human X-chromosome. Cell 47:499–504

Wasmuth JJ, Carlock LR, Smith B, Immken LL (1986) A cell hybrid and recombinant DNA library that facilitate identification of polymorphic loci in the vicinity of the Huntingtion disease gene. Am J Hum Genet 39:397–403

Watterson GA (1974) Models for the logarithmic species abundance distribution. Theo Popul Biol 6:217–250

Weber J, Weith A, Kaiser R, Grzeschik KH, Olek K (1989) Microdissection and microcloning of humam chromosome 7q22–32 region. Somat Cell Mol Genet 16:123–128

Weber G, Greulich KO (1992) Manipulation of cells, organelles and genomes by laser microbeam and optical trap. Int Rev Cytol 131:1–41

Wiegand (Steubing) R, Weber G, Zimmermann K, Monajembashi S, Wolfrum J, Greulich KO (1987) Laser-induced fusion of mammalian cells and plant protoplasts. J Cell Science 88:145–150

Williamson DH, Fennell DJ (1975) The use of florescent DNA-binding agent for detecting and seperating yeast mitochondrial DNA. Methods Cell Biol 12:335–351

Yanagida M, Hiraoka Y, Katsura I (1982) Dynamic behaviors of DNA molecules in solution studied by fluorescence microscopy. Cold Spring Habor Symp Quant Biol 47:177–187

Youngman S, Sarfarazi M, Bucan M, MacDonald M, Smith B, Zimmer M, Gilliam C, Frischauf AM, Wasmuth JJ, Gusella JF, Lehrach H, Harper PS, Shaw DJ (1989) A new DNA marker (D4S90) is located terminally on the short arm of chromosome 4, close to the Huntingtion disease gene. Genomics 5:802–809

Chromosome 12 Aberrations in Human Germ Cell Tumors: A Fluorescence In Situ Hybridization Inventory

R.F. Suijkerbuijk[1], R.J. Sinke[1], L.H.J. Looijenga[2], B.de Jong[3], J.W. Oosterhuis[2], J.J. Cassiman[4], and A. Geurts van Kessel[1]

1 Germ Cell Tumors: General Features

Human germ cell tumors (GCTs) consist of a heterogeneous group of solid neoplasms occurring in the testis, the ovary, and several extragonadal sites (Mostofi and Price 1973; Scully 1979; Gonzales-Crussi 1982). Basically, two entities can be distinguished: (1) tumors composed of neoplastic germ cells, called seminoma in the testis, dysgerminoma in the ovary, and germinoma in extragonadal sites and (2) nonseminomatous GCTs composed of embryonic (embryonal carinoma, immature teratoma, and mature teratoma) and/or extraembryonic tissues (yolk sac tumor and choriocarcinoma; Martineau 1969; Oosterhuis et al. 1990). Testicular GCTs (TGCTs) themselves are also heterogeneous (Atkin 1973; De Jong et al. 1990) and can epidemiologically be subdivided into three groups: infantile tumors, tumors in adults, and tumors in elderly men.

TGCTs exhibit a number of recurring numerical and structural chromosomal abnormalities (Delozier-Blanchet et al. 1986; Castedo et al. 1989c). The most common structural abnormality encountered is the isochromosome 12p, or i(12p) (see Fig. 1). This abnormality constitutes a highly specific and consistent chromosomal marker for TGCTs in adults (Atkin and Baker 1982, 1983, 1985; Gibas et al. 1986; Oosterhuis et al. 1986; Delozier-Blanchet et al. 1985, 1987; Castedo et al. 1989a,b; for review see De Jong et al. 1990). At present, this marker chromosome is found in more than 80% of such TCGTs, regardless of their histology (Castedo et al. 1988; Bosl et al. 1989; De Jong et al. 1990). Furthermore, i(12p) has been observed in dysgerminomas of the ovary (Atkin and Baker 1987; Jenkyn and McCartney 1987; Speleman et al. 1990) and in some extragonadal GCTs (Dal Cin et al. 1989; Chaganti et al. 1989; De Bruin et al. 1994). In the

[1] Department of Human Genetics, University Hospital Nijmegen, P.O. Box 9101, 6500 HB Nijmegen, The Netherlands
[2] Dr. Daniel den Hoed Cancer Center, Rotterdam, The Netherlands
[3] Department for Medical Genetics, University of Groningen, The Netherlands
[4] Center for Human Genetics, University of Leuven, Belgium

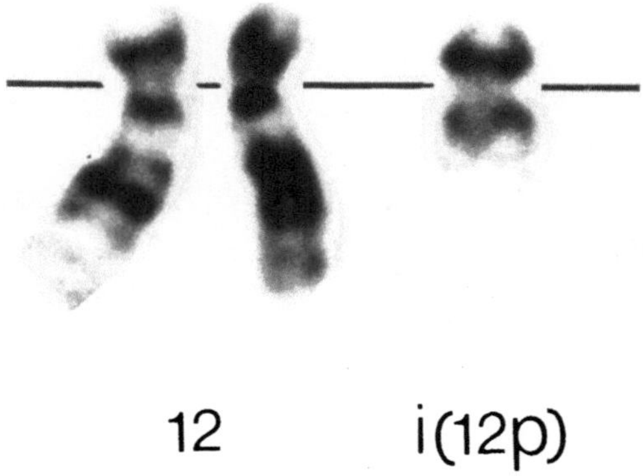

Fig. 1. Giemsa with trypsine-banded partial karyotype showing two copies of a normal chromosome 12 (*left*) and an *i(12p)* (*right*), the standard marker chromosome of germ cell tumors

remaining so-called i(12p)-negative TGCTs, other abnormalities involving chromosome 12 are usually found (Gibas et al. 1986; Castedo et al. 1988).

The occurrence of the i(12p) in GCTs from different anatomical sites points to a common pathogenetic pathway in these tumors, whereby i(12p) plays a predominant role (see below).

2 Testicular Germ Cell Tumors: Pathogenetic Model

The model of pathogenesis of TGCTs as originally proposed by Ewing (1911) and Friedman (1951), according to which seminomas and nonseminomas have a single origin and neoplastic pathway (with seminomas representing an intermediate stage in the development from premalignant carcinoma in situ cells to malignant nonseminoma components), is at present the most favored one. Regarding this model, Vorechovsky and Mazanec (1991) proposed that certain mechanisms are involved in the development of GCTs (and, possibly, in some other tumors also); these are depicted in Fig. 2. The model predicts aneuploidization (due to endoreduplication, cell fusion, or nondisjunction) to be an early step in tumor development (Oosterhuis et al. 1989; Geurts van Kessel et al. 1989). This aneuploidization may conceivably result in genetic instability followed by chromosome loss. In view of this, recent studies dealing with cytogenetic anomalies and ploidy levels of TGCTs (De Jong et al. 1990) showed: (a) gain or retention of whole chromosomes and/or chromosomal segments (namely, from chromosomes 7, 8, 12, and X) and (b) loss of whole chromosomes or chromosomal segments (namely, from chromosome 11, 13, and 18), possibly leading to a selective growth advantage, either by loss of

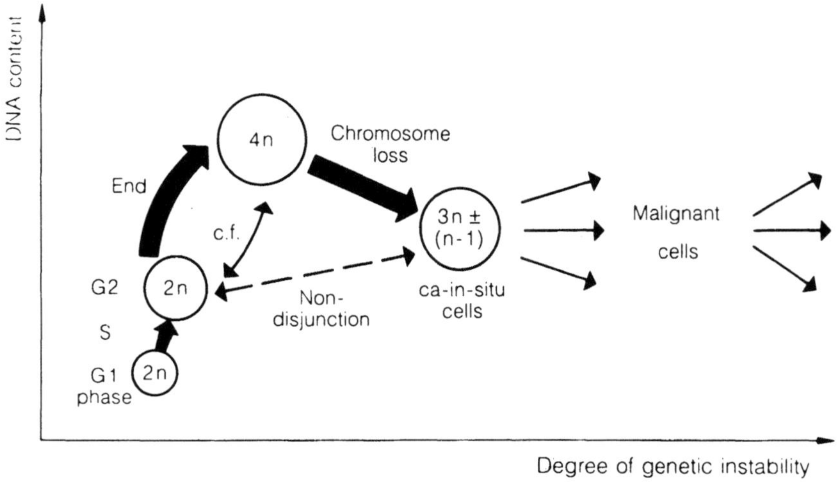

Fig. 2. Possible mechanism contributing to the development of germ cell tumors. End, endoreduplication/endomitosis; c.f., cell fusion; n, haploid set (from Vorechovsky and Mazanec 1991)

tumor supression or loss of the ability for terminal differentiation (see Harris 1986; Sager 1986, for review). Furthermore, the ploidy levels of seminomas appear to be significantly higher than those of nonseminomas (hypertriploid and hypotriploid, respectively), which can be explained as a consequence of a net chromosome loss during tumor progression. The development of structural chromosomal anomalies, in particular the formation of i(12p), may be another result of genetic instability. At present, however, little is known about the exact role of aneuploidization and i(12p) in tumor development. It has been suggested that an increase in i(12p) copy number may be associated with tumor progression and increased resistance to therapy (Delozier-Blanchet et al. 1987; Castedo et al. 1989a; Bosl et al. 1989; Samaniego et al. 1990). This suggestion indicates that its presence and copy number may serve as a diagnostic and prognostic marker in GCTs (Bosl et al. 1989), which emphasizes the importance of a reliable detection of i(12p) in these tumors.

3 Detection of i(12p) in (Testicular) Germ Cell Tumors

Cytogenetic analysis of solid tumors, particularly GCTs, is hampered by the difficulty of obtaining satisfactory numbers of high-quality metaphase spreads which reliably represent the tumorigenic cell population(s). Another problem is the diversity of chromosomal abnormalities that may occur, which obscures the detection of the most relevant recurring chromosomal changes. To circumvent some of these problems, fluorescence in situ hybridization (FISH) strategies have been introduced that allow the detection and localization of specific "target" sequences in microscopic preparations (e.g., interphase nuclei and metaphase chromosomes) by means of fluorescent reporter molecules. With the availability of probes differing in complexity by several orders of magnitude (ranging from ten to 50 nucleotides, as in synthetic nucleotides, to more than 10 million, as in somatic cell hybrids), "target" DNAs varying from single copy sequences to whole chromosomes can be detected more easily and more sensitively than, for example, with previously used radioactive in situ hybridization methods.

Very recently, we reported the application of such FISH strategies to unambiguously demonstrate the presence of a genuine i(12p) chromosome in primary TGCTs and some derived cell lines (Suijkerbuijk et al. 1991, 1992). For that purpose, DNA probes were used to specifically visualize those chromosomal segments that contain chromosome 12-, 12p-, or 12-centromeric sequences (see below). In these studies, other abnormalities involving chromosome 12, which remained unrecognized after application of conventional chromosome-banding techniques and were therefore unexpected, were identified. These results indicated that the FISH method provides a sensitive and reliable method for both confirming cytogenetic data (i.e., i(12p) detection) and obtaining additional karyotypic information in human GCTs.

4 Fluorescence In Situ Hybridization Studies on Germ Cell Tumors and Derived Cell Lines

A diverse group of 12 GCT specimens (six primary tumors and six GCT-derived cell lines, each used as a representative of a distinct testicular GCT; see Table 1) has been examined thoroughly using FISH techniques (Suijkerbuijk et al. 1991, 1992; unpublished observations). To achieve this, the following combinations of probes were used in a bicolor double FISH procedure: (a) biotin-labeled DNA of the chromosome 12-specific library pBS-12 (Collins et al. 1991) and digoxigenin-labeled DNA of the chromosome 12 centromere-specific alphoid sequence pα12H8 (Looijenga et al. 1990) to verify all chromosome 12-derived sequences and (b) biotin-labeled DNA of the monochromosomal human–mouse somatic hybrid cell line M28, containing an i(12p) as its sole human component (Zhang et al. 1989), and digoxigenin-labeled pα12H8 DNA to identify all 12p-derived sequences. In metaphase spreads of each (T)GCT examined, chromosomal segments hybridizing with either digoxigenin-labeled pα12H8 or biotin-labeled pBS-12/M28 were visualized using Texas Red conjugates (red fluorescent staining) or fluorescein isothiocyanate (FITC) conjugates (yellow-green staining), respectively. For contrast, chromosomes were

Table 1. Summary of all germ cell tumors included in the present study, their origin and cytogenetic (chromosome 12) constitution as revealed by GTG banding and fluorescence in situ hybridization (FISH)

Tumor code	Modal chromosome number	Tumor type	References	Chromosome 12 constitution revealed by GTG banding and FISH					
				Chromosome 12		i(12p)		Markers	
				GTG	FISH	GTG	FISH	GTG	FISH
Ma-90	48	DGCT; GO/DYS/IT/TM/YS	Hamers et al. 1991	2	2	1[1]	1	–	–
KDK[2]	57–60	TGCT; EC	Suijkerbuijk et al. 1992	3	3	3–4[1]	3 (2)[3]	0-mar1[1]	mar1 (mar1–mar2)[3]
Germa-1[2]	54–57	TGCT; EC	Hofmann et al. 1989	2	2	1–2[1]	1	0-mar1[1]	mar1–mar2
NT2/D1[2]	57–60	TGCT; EC	Thompson et al. 1984	3	3	2	1[4]	–	mar1
Sp-90	49	OGCT; TM	Speleman et al. 1992	2	2	1[1]	1	–	–
T2219	64	IGCT; EC/TM/CC	De Bruin et al. 1994	3	3	2[1]	2	–	–
Scha-1[2]	61–63	TGCT; EC	Andrews et al. 1987	2	2	–	–	–	mar1-mar4
T87-1004	77–79	MGCT[1]; TM	Oosterhuis et al.[5]	3	3	–	–	–	–
TT34DA-tuc[2]	106	TGCT; SE/TI	Parrington et al.[5]	5	5	–	–	mar1[1]	mar1–mar3
TT34DA-tuj[2]	64	TGCT; SE/TI	Parrington et al.[5]	3	3	–	–	mar1[1]	mar1–mar3
ST-89-07436	56	TGCT; EC/TM	Suijkerbuijk et al. 1993	2	2	–	–	mar1[1]	mar1
ST-90-13401	58	TGCT; SE/EC/TI	Suijkerbuijk et al. 1993	2	2	–	–	mar1[1]	mar1–mar2

GTG, Giemsa banding with trypsine; DGCT, dysgenetic germ cell tumor; TGCT, testicular germ cell tumor; OGCT, ovarian germ cell tumor; IGCT, intracranial germ cell tumor; MGCT, mediastinal germ cell tumor; GO, gonadoblastoma; DYS, dysgerminoma; SE, seminoma; EC, embryonal teratoma; IT, immature teratoma; TM, mature teratoma; TI, teratoma intermediate; YS, yolk sac tumor; CC, choriocarcinoma.

[1] Questionable identity of (one of) the respective subject(s).

[2] Tumor-derived cell line.

[3] Found in less than 25% of the cells.

[4] Not including the i(12p)+ marker chromosome, which has been denoted as mar1 in this table.

[5] Unpublished observations.

counterstained with the DNA-specific dye DAPI (blue). The FISH results thus obtained were evaluated using the next criteria: only those copies of chromosome 12 which were of normal size and exhibited a normal centromere index and 4,6-diamidino-2-phenylindole (DAPI) staining pattern were considered to be normal (although minor structural aberrations such as point and insertional mutations, small deletions, inversions, or amplifications may still be present). All other chromosomes containing chromosome 12-derived sequences were designated as marker chromosomes (mar in Table 1; m in Figs. 3, 4). In addition, the 12 material-containing chromosomes were compared with their GTG (Giemsa banding with trypsine) banded counterparts. In this way, the presence of any karyotypically described chromosome 12 and 12-derived marker chromosome(s), especially i(12p), in a given tumor could be verified. Furthermore, those chromosome 12-derived abnormalities that had remained unidentified after conventional chromosome banding analyses could be visualized.

4.1 Results of i(12p)-Positive Germ Cell Tumors and Derived Cell Lines

After application of FISH, an easy and clearcut identification of all normal copies of chromosome 12 and 12-derived marker chromosomes present in each GCT was established (see Fig. 3). The identification of 12p-derived sequences in particular revealed more detailed information and was, therefore, more powerful than conventional GTG-banding analyses: in three GCTs (Ma-90, Sp-90, and T2219), the identity of presumed i(12p) chromosomes could be confirmed, while in another three GCTs (KDK, Germa-1, and NT2/D1), a clearcut discrimination between genuine i(12p) chromosomes and other, metacentric chromosome 12-derived marker chromosomes which strongly resembled i(12p) chromosomes and were similarly sized (the identities of which had remained questionary up till then) was possible after FISH only (see Table 1). As a consequence, it appears that FISH methods allow a more accurate estimation of i(12p) copy numbers present in GCTs than do conventional cytogenetic analyses. Moreover, they enable the identification of previously unrecognized chromosome 12-derived abnormalities. Interestingly, this latter group of abnormalities all show involvement of the short arm of chromosome 12 (see, for example, Fig. 3), indicating that i(12p) may not be the only relevant 12p-derived anomaly found in GCTs (Suijkerbuijk et al. 1991, 1992).

4.2 Results of i(12p)-Negative Germ Cell Tumors and Derived Cell Lines

FISH analyses of three i(12p)-negative primary GCTs and three i(12p)-negative GCT-derived cell lines revealed that the observations (i.e., struc-

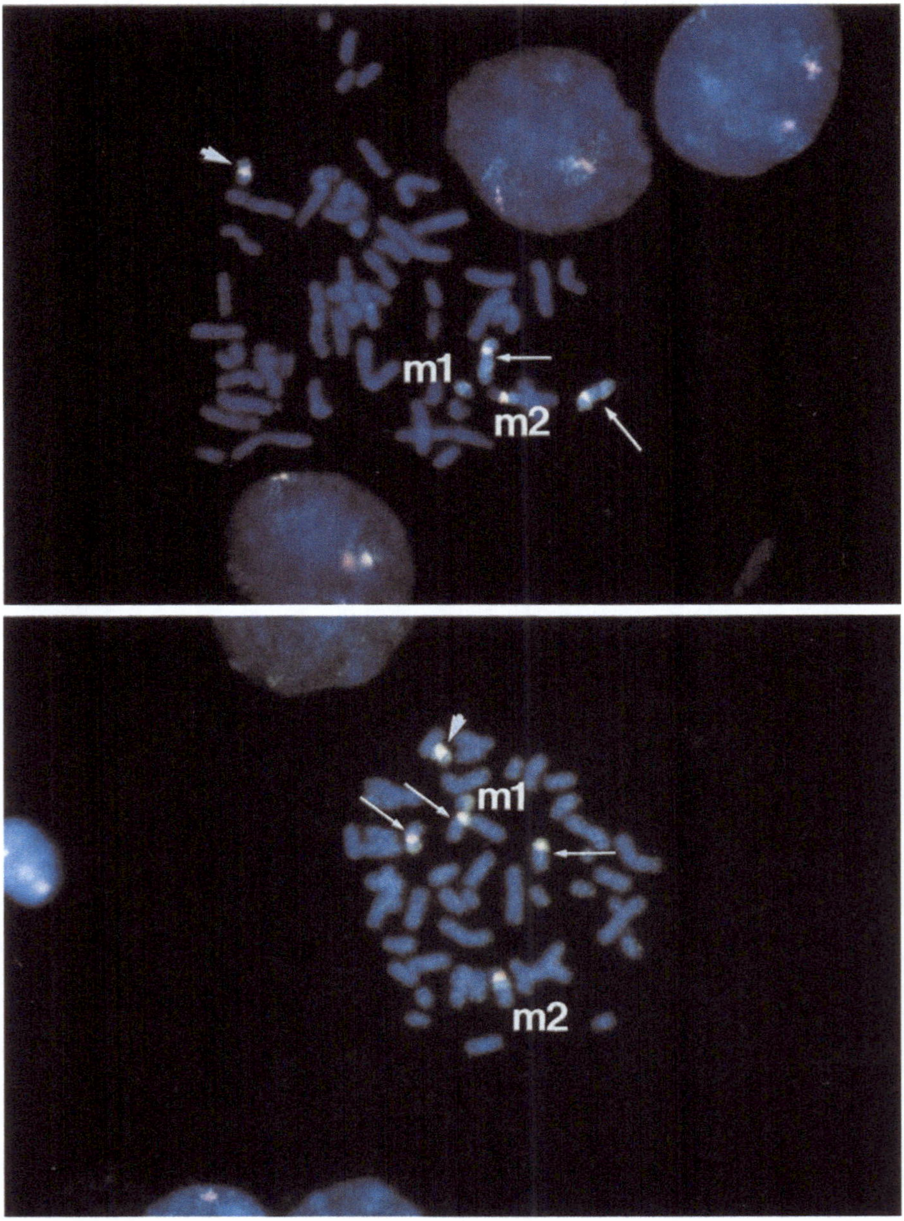

Fig. 3. Results of bicolor double fluorescence in situ hybridization using either pα12H8 (*red*) and pBS-12 (*yellow; upper*) or p*α12H8 (red*) and M28 (*yellow; lower*) as probes on metaphase spreads of the i(12p)-positive testicular germ cell tumor-derived cell line Germa-1. Chromosomes are counterstained 4,6-diamidino-2-phenylindole (*blue*). Copies of a normal chromosome 12, i(12p), and 12-derived marker chromosomes are indicated by *arrows*, *arrowheads*, and *m1-2*, respectively

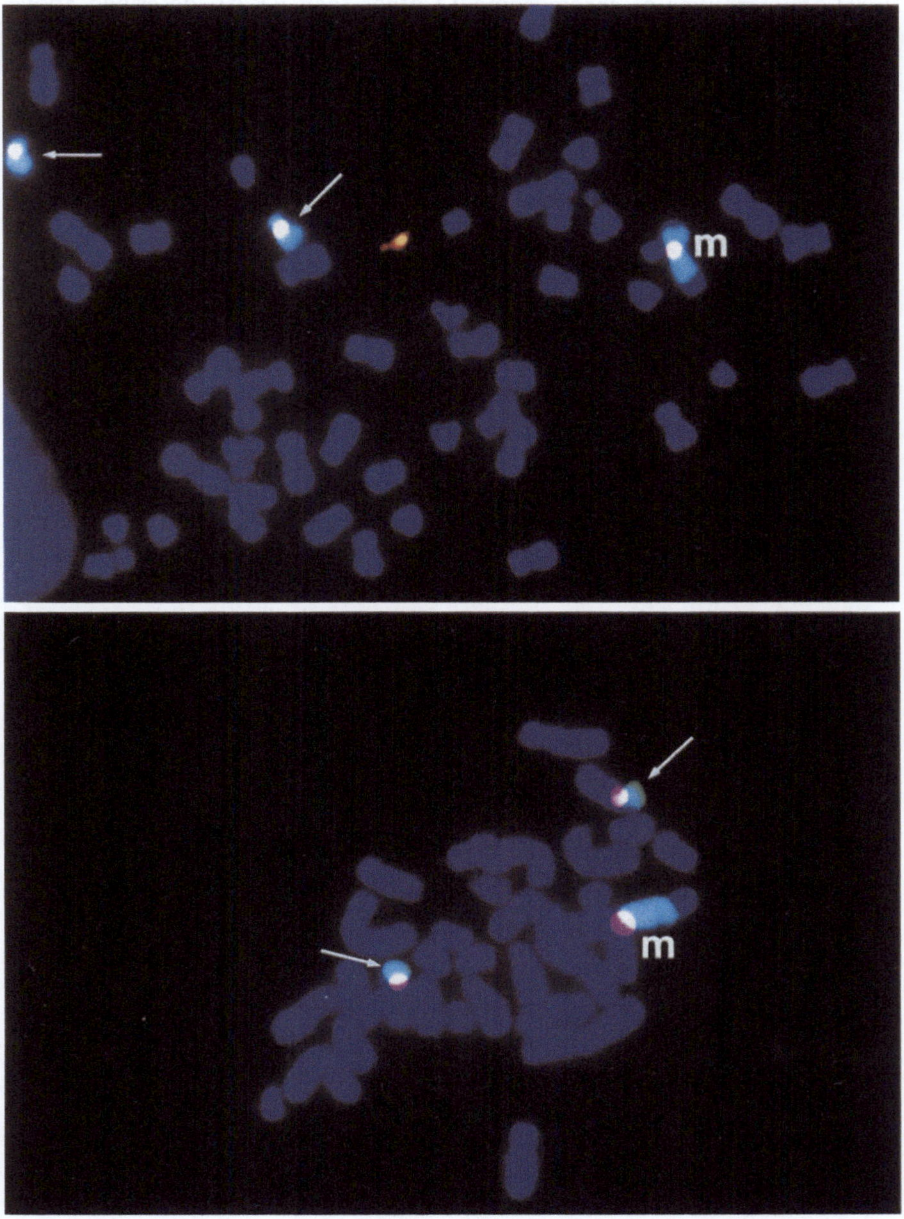

Fig. 4. Fluorescence in situ hybridization patterns of either pα12H8 (*red*) and pBS-12 (*yellow; upper*) or pα12H8 (*red*) and M28 (*yellow; lower*) as probe mixtures on metaphase spreads of the i(12p)-negative testicular germ cell tumor ST-89-07436. Chromosomes are counterstained with 4,6-diamidino-2-phenylindole. Copies of chromosome 12 are indicated with *arrows*; the chromosome 12-derived marker chromosome is depicted by a *m*

tural abnormalities of chromosome 12 other than i(12p)) described above for i(12p)-positive GCTs could be extended further. With the exception of one extragonadal GCT, all GCTs and GCT-derived cell lines (which were all of the testicular subtype) showed structural abnormalities of chromosome 12 (see Fig. 4 and Table 1). Careful examination of the FISH results revealed that these abnormalities involved 12p-derived sequences, most of which had, again, remained undetected by previous conventional cytogenetic banding (e.g., GTG banding) analyses (see Table 1). All 12p abnormalities resulted either from a translocation, an inversion, or a multiplication event and included nearly the entire 12p arm (Suijkerbuijk et al. 1992; unpublished observations). It seems likely that the appearance of such rearrangements must have been preceded by chromosomal breaks at or near the telomeric and/or centromeric regions, because the rearrangements invariably involved the 12p11–13 region. Thus, the presence of structural abnormalities of chromosome 12 other than i(12p) appears to be a consistent, although less perceptible, phenomenon in both i(12p)-positive and -negative GCTs. In analogy to i(12p), these abnormalities may contribute to the process of GCT development and progression. Similarly, Castedo et al. (1988) suggested that such abnormalities in i(12p)-negative TGCTs might, in combination with other structural and numerical abnormalities, have the same effect as i(12p) itself. Clearly, more profound and detailed clinical as well as cytogenetic and molecular studies on both i(12p)-positive and -negative TGCTs are needed to clarify the exact pathogenetic significance of these chromosome 12 abnormalities in GCTs.

5 New Fluorescence In Situ Hybridization Strategies to Study Structural Abnormalities of Chromosoms 12 in Germ Cell Tumors in More Detail

Our FISH data suggest that besides i(12p) also other chromosome 12 aberrations are a common phenomenon in both i(12p)-positive and -negative GCTs. At present one can only speculate about the exact mechanism(s) by which i(12p) and other 12p abnormalities may contribute to the oncogenic process: its appearance may disturb a fine-tuned and narrow-defined balance between 12p sequences and sequences derived from 12q and/or other chromosomes. Alternatively, gene(s) located near putative breakpoint cluster regions on 12p may be involved (i.e., deregulated; Suijkerbuijk et al. 1992). Rearrangements of chromosome 12, as described above, resulting in chromosomal aberrations other than i(12p) may be of particular value in defining such region(s) on chromosome 12.

Therefore, we have employed an irradiation and fusion procedure to generate somatic cell hybrids containing various fragments of the short arm of human chromosome 12 that may be used as a resource for the devel-

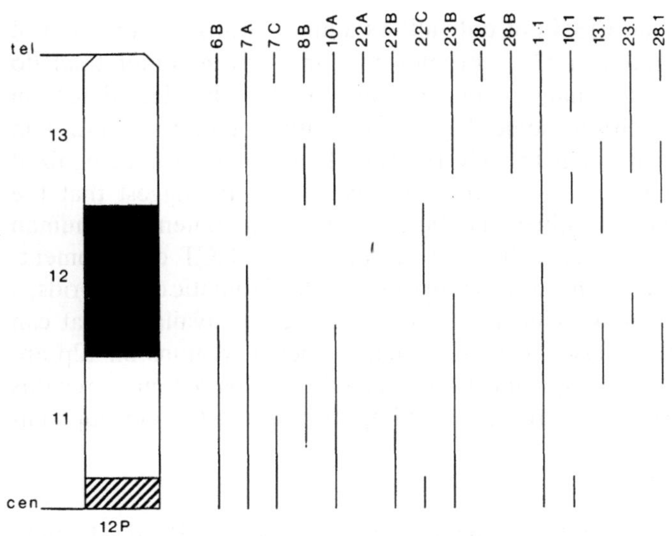

Fig. 5. The human chromosome 12p constituents of the radiation-reduced WgM hybrids resulting from combining both Southern blot analyses and chromosome painting (from Sinke et al. 1992)

opment of new FISH (and other) probes from well-defined regions of 12p (Sinke et al. 1992). For this purpose, the 12p-only somatic cell hybrid cell line M28 (Zhang et al. 1989) was used. With the irradiation and fusion technique first described by Goss and Harris (1975), 17 cell lines containing detectable amounts of human material were selected. Detaild characterization of these hybrids by Southern blot and FISH analyses resulted in a radiation-reduced hybrid panel encompassing the entire chromosome 12p arm (see Fig. 5). This panel provides a valuable resource for the rapid isolation of region-specific DNA probes that can be applied to the characterization of cytogenetic anomalies in GCTs. Moreover, since amplified DNA (by interspersed element-mediated polymerase chain reaction methods; Lengauer et al. 1990; Lichter et al. 1990) of these hybrids themselves can also directly be used as region-specific probes for FISH purposes (Sinke et al. 1992), this panel will be helpful in the identification of a region or regions on 12p that are nonrandomly involved in chromosome 12 aberrations that occur in human GCTs.

6 Conclusions and Future Perspectives

We have demonstrated that an easy and reliable visualization of chromosome 12 aberrations in human TGCTs and other GCTs can be achieved by FISH techniques using appropriate probes (e.g., pα12H8, M28, and pBS-12). Our results have shown that FISH not only allows a more accurate estimation of copy numbers of genuine i(12p) chromosomes in a given tumor than do conventional cytogenetic analyses, but may also lead to the clear detection and identification of chromosome 12p abnormalities other than i(12p) in both i(12p)-positive and -negative GCTs which had remained unrecognized by cytogenetic means only. As a consequence, our data suggest that the occurrence of such abnormalities may be a common phenomenon in human GCTs and, as such, may contribute to the process of GCT development. With the availability of a panel of 12p radiation-reduced somatic cell hybrids, a new collection of region-specific 12p-probes has become available that can be applied for FISH analyses to verify whether specific regions on 12p are nonrandomly involved. If so, it must be assumed that this region or regions harbor a gene or genes that are of major importance in GCT development and progression.

Acknowledgements. The authors thank Drs. D. Smeets, B. de Leeuw, and H.-H. Ropers for advice and critical reading of the manuscript; H. Janssen, J. van Gaal, W. van de Klundert, A. Siepman, and F. Stellink for excellent technical support; and M. Ariaans and H. Mannuputty for secretarial assistance. Furthermore, we are very grateful to Drs. J.M. Parrington, H. Walt, A.M. Meloni, and A.A. Sandberg for their kind gift of the tumor material KDK and TT34DA-tuc/tuj, Germa-1, ST-89-07436, and ST-90-13401, respectively. This work was supported by grant GUKC 90-17 from the Netherland Cancer Society (Koningin Wilhelmina Fonds) and grant Ro 389/15-4 (to H.-H. Ropers) from the Deutsche Forschungsgemeinschaft.

References

Andrews PW, Oosterhuis JW, Damjanov I (1987) Cell lines from human germ cell tumors. In: Robertson E (ed) Teratocarcinomas and embryonal stem cell lines. IRL Press, Oxford, pp 207–248

Atkin NB (1973) High chromosome numbers of seminomata and malignant teratoma of the testis: a review of data on 103 tumours. Br J Cancer 28:275–279

Atkin NB, Baker MC (1982) Specific chromosome change, i(12p) in testicular tumours? Lancet 2:1349

Atkin NB, Baker MC (1983) i(12p): specific chromosomal marker in seminoma an malignant teratoma of the testis? Cancer Genet Cytogenet 10:199–204

Atkin NB, Baker MC (1985) Chromosome analysis of three seminomas. Cancer Genet Cytogenet 17:315–323

Atkin NB, Baker MC (1987) Abnormal chromosomes involving small metacentrics in 14 ovarian cancers. Cancer Genet Cytogenet 26:355–361

Bosl GJ, Dmitrovsky E, Reuter VE, Samaniego F, Rodriquez E, Geller NL, Chaganti RSK (1989) Isochromosome of chromosome 12: clinically useful marker for male germ cell tumors. J Natl Cancer Inst 81:1874–1878

Castedo SMMJ, de Jong B, Oosterhuis JW, Seruca R, Idenburg V, Buist J, Sleijfer DTh (1988) i(12p) negative testicular germ cell tumors. A different group? Cancer Genet Cytogenet 35:171–178

Castedo SMMJ, de Jong B, Oosterhuis JW, Idenburg V, Seruca R, Buist J, te Meerman G, Schraffordt Koops H, Sleijfer DTh (1989a) Chromosomal changes in mature residual teratomas following polychemotherapy. Cancer Res 49:672–676

Castedo SMMJ, de Jong B, Oosterhuis JW, Seruca R, Idenburg V, Dam A, te Meerman G, Schraffordt Koops H, Sleijfer DTh (1989b) Chromosomal changes in primary testicular nonseminomas. Cancer Res 49:5696–5701

Castedo SMMJ, de Jong B, Oosterhuis JW, Seruca R, te Meerman G, Dam A, Schraffordt Koops H (1989c) Cytogenetical analysis of ten human seminomas, two of them lacking i(12p). Cancer Res 49:439–443

Chaganti RSK, Ladanyi M, Samaniego F, Offit K, Reuter VE, Jhanwar SC, Bosl GJ (1989) Leukemic differentiation of a mediastinal germ cell tumor. Genes Chromosomes Cancer 1:83–87

Collins C, Kuo WL, Segraves R, Fuscoe J, Pinkel D, Gray JW (1991) Construction and characterization of plasmid libraries enriched in sequences from single human chromosomes. Genomics 11:997–1006

Dal Cin P, Drochmans A, Moerman P, Van den Berghe H (1989) Isochromosome 12p in mediastinal germ cell tumor. Cancer Genet Cytogenet 42:243–251

De Bruin TWA, Slater RM, Defferreri R, Geurts van Kessel A, Suijerbuijk RF, Jansen G, de Jong B, Oosterhuis JW (1994) Isochromosome 12p positive intracranial germ cell tumor. Cancer Res 54:1542–1544

De Jong B, Oosterhuis JW, Castedo SMMJ, Vos A, te Meerman GJ (1990) Pathogenesis of adult testicular germ cell tumors. A cytogenetic mode (10th anniversary article). Cancer Genet Cytogenet 48:143–167

DeLozier-Blanchet CD, Engel H, Walt H (1985) Isochromosome 12p in malignant testicular tumors. Cancer Genet Cytogenet 15:375–376

DeLozier-Blanchet CD, Walt H, Engel E (1986) Cytogenetic studies of human testicular germ cell tumours. Int J Androl 10:69–77

DeLozier-Blanchet CD, Walt H, Engel E, Vaugnat P (1987) Cytogenetic studies of human testicular germ cell tumours. Int J Androl 10:69–78

Ewing J (1911) Teratoma testis and its derivates. Surg Gynecol Obstet 12:230–261

Friedman NB (1951) The comparative morphogenesis of extragenital and gonadal teratoid tumors. Cancer 4:265–275

Geurts van Kessel A, van Drunen E, de Jong B, Oosterhuis JW, Langeveld A, Mulder MP (1989) Chromosome 12q heterozygosity is retained in i(12p)-positive testicular germ cell tumor cells. Cancer Genet Cytogenet 40:129–134

Gibas Z, Prout GR, Pontes JE, Sandberg AA (1986) Chromosome changes in germ cell tumors of the testis. Cancer Genet Cytogenet 19:245–252

Gonzales-Crussi F (1982) Extragonadal teratomas. AFIP atlas of tumor pathology, second series, fascicle 18, Washington DC

Goss SJ, Harris H (1975) New method for mapping genes in human chromosomes. Nature 255:1445–1458

Hamers A, de Jong B, Suijkerbuijk RF, Geurts van Kessel A, Oosterhuis JW, van Echten J, Evers J, Bosman F (1991) A 46,XY female with mixed gonadal dysgenesis and a 48,XY,+7,+i(12p) chromosome pattern in a primary gonadal tumor. Cancer Genet Cytogenet 57:219–224

Harris H (1986) The genetic analysis of malignancy. J Cell Sci [Suppl] 4:431–444

Hofmann MC, Jetsch W, Brecher J, Walt H (1989) Alkaline phosphatase isoenzymes in human testicular germ cell tumors, their precancerous stage, and three related cell lines. Cancer Res 49:4696–4700

Jenkyn DJ, McCartney AJ (1987) A chromosome study of three ovarian tumors. Cancer Genet Cytogenet 26:327–337

Lengauer C, Riehtman H, Cremer T (1990) Painting of human chromosome with probes generated from hybrid cell lines by PCR with Alu and L1 primers. Hum Genet 86:1–6
Lichter P, Ledbetter SA, Ledbetter DH, Ward DC (1990) Fluorescence in situ hybridization with Alu and L1 polymerase chain reaction probes for rapid characterization of human chromosomes in hybrid cell lines. Proc Natl Acad Sci USA 87:6634–6638
Looijenga LHJ, Smits VTHBM, Wessels JW, Mollevanger P, Oosterhuis JW, Cornelisse CJ, Devilee P (1990) Localization and polymorphism of a chromosome 12-specific alpha satellite DNA sequence. Cytogenet Cell Genet 53:216–218
Martineau M (1969) Chromosomes in human testicular tumours. J Pathol 99:271–282
Mostofi FK, Price EB (1973) Tumors of the male genital system. AFIP atlas of tumor pathology, second series, fascicle 8, Washington DC
Oosterhuis JW, de Jong B, Cornelisse CJ, Molenaar WM,Meiring A, Idenburg VJS,Schraffordt Koops H, Sleijfer DT (1986) Karyotyping and DNA flow cytometry of mature residual teratoma after intensive chemotherapy of disseminated nonseminomatous germ cell tumor of the testis: a report of two cases. Cancer Genet Cytogenet 22:149–157
Oosterhuis JW,Castedo SMMJ, de Jong B, Cornelisse CJ, Dam A, Sleijfer DT, Schraffordt Koops H (1989) primary germ cell tumors of the testis. Pathogenetic and clinical relevance. Lab Invest 60:14–21
Oosterhuis JW, Castedo SMMJ, de Jong B (1990) Cytogenetics, ploidy and differentiation of human testicular, ovarian and extragonadal germ cell tumours. Cancer Surveys 9:321–332
Sager R (1986) Genetic suppression of tumor formation: a new frontier in cancer research. Cancer Res 46:1573–1580
Samaniego F, Rodriquez E, Houldsworth J, Murty VVVS, Ladanyi M, Pele KP, Chen Q, Dmitrovsky E, Geller NL, Reuter V, Jhanwar SC, Bosl GJ, Chaganti RSK (1990) Cytogenetic and molecular analysis of human male germ cell tumors: chromosome 12 abnormalities and gene amplification. Genes Chromosomes Cancer 1:289–300
Scully RE (1979) Tumors of the ovary and maldeveloped gonads. AFIP atlas of tumor pathology, second series, fascicle 16, Washington DC, p 16
Sinke RJ, Suijkerbuijk RF, Herbergs J, Janssen H, Cassiman JJ, Geurts van Kessel A (1992) Generation of a panel of somatic cell hybrids containing fragments of human chromosome 12p by X-ray irradiation and cell fusion. Genomics 12:206–213
Speleman F, De Potter, Dal Cin P, Mangelschots K, Ingelaere H, Laureys G, Benoit Y, Leroy J, Van den Berghe H (1990) i(12p) in malignant ovarian tumor. Cancer Genet Cytogenet 45:49–53
Speleman F, Laureys G, Benoit Y, Cuvelier C, Suijkerbuijk RF, de Jong B (1992) i(12p) in a near diploid mature ovarian teratoma. Cancer Genet Cytogenet 60:216
Suijkerbuijk RF, van der Veen AY, van Echten J, Buys CHCM, de Jong B, Oosterhuis JW, Warburton DA, Cassiman JJ, Schonk D, Geurts van Kessel A (1991) Demonstration of the genuine iso(12p) character of the standard marker chromosome of testicular germ cell tumors and identification of further chromosome 12 aberrations by competitive in situ hybridization. Am J Hum Genet 48:269–273
Suijkerbuijk RF, Looijenga L, de Jong B, Oosterhuis JW, Cassiman JJ, Geurts van Kessel (1992) Verification of isochromosome 12p and identification of other chromosome 12 aberrations in human germ cell tumors by bicolor double fluorescence in situ hybridization. Cancer Genet Cytogenet 63:8–16
Suijkerbuijk RF, Sinke RJ, Meloni AM, Parrington JM, van Echten J, de Jong B, Oosterhuis JW, Sandberg AA, Geurts van Kessel A (1993) Overrepresentation of chromosome 12p sequences and karyotypic evolution in i(12p)-negative testicular germ cell tumors revealed by fluorescence in situ hybridization. Cancer Genet Cytogenet 70:85–93
Thompson S, Stern PL, Webb M, Walsh FS, Engstrom W, Evans EP, Shi WK, Hopkins B, Graham CF (1984) Cloned human teratoma cells differentiate into neuron-like cells and other cell types in retinoic acid. J Cell Sci 72:37–64
Vorechovsky I, Mazanec K (1991) Cytogenetic investigation of gonadal carcinomas in situ syndromes of abnormal differentiation. In: Oosterhuis JW, Walt H, Damjanov I (eds)

Pathobiology of human germ cell neoplasia. Springer, Berlin Heidelberg New York, pp 45–49 (Recent results in cancer research, vol 123)
Zhang J, Marynen P, Devriendt K, Fryns JP, Van den Berghe H, Cassiman JJ (1989) Molecular analysis of the isochromosome 12p in the Pallister-Killian syndrome. Construction of a mouse-human hybrid cell containing an i(12p) as the sole human chromosome. Hum Genet 83:359–363

Development and Characterization of Cell Lines from a Myxoid Liposarcoma with t(12;16)(q13;p11.2) and Trisomy 8

F.P.M. Schoenmakers[1], B. Kazmierczak[2], P.F.J. Kools[1], R. Mols[1], P. Dal Cin[1], J. Bullerdiek[2], H. Van den Berghe[1], and W.J.M. Van de Ven[1]

1 Introduction

The present state of knowledge about chromosome aberrations and cancer clearly emphasizes the important and direct contributions of molecular cytogenetics to the more detailed description and understanding of oncogenesis (Sandberg 1990; Mitelman 1991; Solomon et al. 1991). Although fundamental genetic analysis of leukaemias and lymphomas has advanced more rapidly in the past than that of solid tumours, extensive knowledge about solid tumours has recently also been obtained due to improvements in cytogenetic technology. Many solid tumours appear to exhibit often an irregular and complex pattern of chromosome aberrations. However, consistent chromosome aberrations have also been observed (Sandberg 1990; Mitelman 1991). Region q13–15 of chromosome 12, for instance, is frequently rearranged. Analysis of human myxoid liposarcomas has generally revealed a consistent reciprocal translocation between chromosomes 12 and 16, t(12;16)(q13;p11), or a three-way translocation also involving chromosome 1 (Turc-Carel et al. 1986a; Ohjimi et al. 1992). Interestingly, in cytogenetically abnormal lipomas the same cytogenetic region of chromosome 12 is frequently implicated in translocations; in contrast to myxoid liposarcomas, however, multiple chromosomes are found as translocation partner (Turc-Carel et al. 1986b; Sreekantaiah et al. 1991). This is also the case for cytogenetically abnormal uterine leiomyomas (Pandis et al. 1991) as well as cytogenetically abnormal pleomorphic adenomas of the salivary gland (Sandros et al. 1990). Although these studies have indicated that the same cytogenetic region of chromosome 12 is often involved in chromosome translocations in these four solid tumours, the precise nature of the chromosome 12 breakpoints in the various tumours is still not known, nor are the genes that are affected directly by the translocations.

Molecular studies are required to resolve these matters, and they are likely to provide further insights into the underlying mechanisms of tumour

[1] Center for Human Genetics, University of Leuven, Herestraat 49, 3000 Leuven, Belgium
[2] Zentrum für Humangenetik, Universität Bremen, ZHG, Leobenerstrasse, 28359 Bremen, Germany

development. Of crucial importance for such studies is the availability of appropriate tumour cell lines and somatic cell hybrids that have retained the derivative chromosome(s) of interest in stable form, since large amounts of cells with the particular chromosomal anomalies can than be obtained. In recent studies, cell lines with an increased growth potential were obtained from primary solid tumours by transfecting cells of these with DNA constructs harbouring the "early region" of the genome of the SV40 virus (Kazmierczak et al. 1990; Stern et al. 1991). Cell lines were obtained from pleomorphic adenomas of the salivary gland (Kazmierczak et al. 1990), uterine leiomyomas (Stern et al. 1991), and lipomas (Belge et al. 1992). Here we report about the development of tumour cell line LIS-3/SV40 from a myxoid liposarcoma with a t(12;16)(q13;p11.2) and a trisomy 8; furthermore, a panel of man/mouse somatic cell hybrids was generated using cells from the newly developed LIS-3/SV40 cell line, and hybrids were selected from these which all contained der(16) as the sole derivative chromosome and at the same time lacked human chromosomes 12 and 16. Characteristics of cell hybrid LIS-3/SV40/A9-B4 are described.

2 Materials and Methods

2.1 Production of Somatic Cell Hybrids

To generate man/mouse somatic cell hybrids, mouse A9 cells, which are $APRT^-$ adenine phosphoribosyltransferase negative ($APRT^-$) and hypoxanthine phosphoribosyltransferase negative ($HPRT^-$) were fused to human myxoid liposarcoma LIS-3/SV40 cells with a t(12;16)(q13;p11.2) and a trisomy 8. In cell fusion experiments, 3×10^7 LIS-3/SV40 cells were mixed with 8×10^6 mouse A9 cells and washed twice in Dulbecco's modified eagle's medium (DME) supplemented with 10% fetal bovine serum and incubated in the same medium for 24 h. Subsequently, cells were washed twice in Ca^{2+}-free MAM medium (containing per litre 6.7 g NaCl, 1.0 g glucose, 0.4 g KCl, 0.035 g $MgCl_2$, and 2.383 g HEPES, 10 mM, pH 7.5) and incubated with polyethylene glycol (PEG-1500, Boehringer, Mannheim) for 90 seconds (Pontecorvo 1975). Upon removal of PEG-1500, cells were washed three times (10 s) with Ca^{2+}-free MAM of 37°C, and thereafter cells were incubated for 30 min in Ca^{2+}-free MAM at 37°C. Subsequently, the medium was removed and replaced by selective AOA medium (consisting of DME-F12 medium supplemented with 10% fetal bovine serum, 0.05 mM adenine, 0.05 mM ouabain, and 0.01 mM azaserine), and cells were grown for 24 h at 37°C. After this 24-h incubation period, cells were treated with trypsin and explanted at high dilution. Upon prolonged incubation of the cells in selective AOA, colonies of proliferating somatic cell hybrids were isolated and used for further studies.

2.2 Chromosome Preparations and Fluorescence In Situ Hybridization Analysis

Metaphase spreads from LIS-3/SV40 and LIS-3/SV40/A9-B4 cells were prepared from cultured cells harvested 48 h after passaging. Colcemid (25 ng/ml) was added 2 h before harvesting. After hypotonic shock in 75 m*M* KCl, cells were fixed using methanol: acetic acid (3:1), and chromosome spreads were prepared on glass slides according to routine procedures. Subsequently, chromosome preparations were treated with RNase A (0.1 mg/ml in 2 × SSC, 1 h at 37°C), 0.01% pepsin (in 0.01 *M* HCl, 10 min at 37°C) and fixed in 1% formaldehyde/50 m*M* $MgCl_2$/PBS before dehydration through an ethanol series. Subsequently, the chromosomal DNA was denatured in 70% formamide/2 × SSC/50 m*M* sodium phosphate (pH 7) under a coverslip for 2–3 min on a hot plate (80°C). Renaturation was prevented by subsequent washing in cold 70% ethanol, followed by dehydration.

In chromosome 12 painting experiments, pBS12 DNA (Collins et al. 1991), which consisted of the genomic DNA of flow-sorted chromosomes 12 cloned into Bluescribe (Stratagene), was used as probe. pBS12 DNA was labelled with biotin-11-dUTP (Boehringer, Mannheim) using the BRL Nick Translation system (Life Technologies). In chromosome 16 painting experiments, a pool of 166 different cosmids from chromosome 16 was used. DNA of the 166 cosmids was labelled with biotin-11-dUTP using the protocol of Kievits et al. (1990).

Hybridization Conditions. Regarding the pool of chromosome 12 specific probes, labelled pBS12 DNA (5 ng/μl) and a 20-fold excess of unlabelled human Cot1 DNA (BRL, Life Technologies) were dissolved in hybridization solution (50% formamide/2 × SSC/50 m*M* sodium phosphate/10% dextran sulphate pH 7) and, upon denaturation at 75°C for 5 min, allowed to reanneal for 15 min at 37°C. With respect to the pool of chromosome 16 cosmids, different conditions were used; labelled DNA with a 1000-fold excess of human placenta DNA (Sigma) was ethanol precipitated and dissolved in hybridization solution; after denaturation, the probe mixture was allowed to reanneal for 90 min. Denatured chromosomes on prewarmed slides (37°C) were hybridized under a coverslip; coverslips were sealed with rubber cement. Hybridization took place overnight at 37°C in a humid chamber. Following hybridization, chromosome preparations were washed three times for 5 min in 50% formamide/2 × SSC pH 7 (45°C) and three times for 5 min in 2 × SSC (45°C). To reduce non-specific hybridization, the chromosome preparations hybridized with the pool of cosmids from chromosome 16 were washed three times for 5 min in 2 × SSC at 60°C. Thereafter, preparations were incubated with 5% non-fat dry milk (Lucerne) in 4 × SSC (NFDM buffer). Subsequently, bound probe was visualized by serial incubations at room temperature for 20 min using avidin^{FITC} (vector) and goat-α-avidin^{biotin} (vector). Every incubation was followed by three washes of 5 min in 4 × SSC/0.05% Tween-20. In total, three layers of avidin^{FITC} were applied. Results were evaluated on a Zeiss Axiophot fluorescence microscope and recorded on Scotch (3 M) film (640 asa).

G Banding. After in situ hybridization analysis, chromosome preparations were washed first three times for 10 min in 2 × SSC and subsequently once for 5 min in 70% ethanol; finally, the chromosomes were dehydrated in an ethanol series. Chromosomes were banded using a slightly modified version of the Wright staining protocol (De la Maza and Sanchez 1976). Briefly, chromosome preparations on slides were incubated for 20 min in 2 × SSC at 60°C, rinsed shortly in demineralized water and dried. Staining of the chromosomes was achieved by incubating the preparations for 3 min in Wright's staining solution [3 ml phosphate buffer (Na, K phosphate, 60 m*M*, pH 6.8) mixed with 1 ml Wright's solution; Wright's solution = 0.25% (w/v) Wright's stain in methanol, filtered and incubated 2–5 days at 37°C].

3 Results and Discussion

3.1 Development of Cell Line LIS-3/SV40 from a Primary Human Myxoid Liposarcoma with a t(12;16)(q13;p11.2) and Trisomy 8

Primary tumour cells were obtained from a male patient 68 years old from whom a tumour (5 × 5 × 4 cm) was ressected from the right groin; upon histopathological examination this tumour was diagnosed as a myxoid liposarcoma. To culture the myxoid liposarcoma cells, a sample of the tumour was carefully minced and subsequently treated with 0.8% collagenase for 4 h. The cell suspension was centrifuged and resuspended in TC199 medium with Earle's salts supplemented with 20% fetal bovine serum and the antibiotics penicillin (200 IU/ml) and streptomycin (200 μg/ml). Some of these cells were used for karyotypic analysis, and of the 33 metaphases analysed 12 metaphases had 46,XY,t(12;16)(q13;p11.2) and 21 metaphases 47,XY,+8 t(12;16)(q13;p11.2). Subsequently, the remainder of the tumour cells were transfected with 10 μg pSV40-ΔN DNA, which consisted of plasmid pAT153 containing in its *Bam*HI site genetic sequences of the SV40 early region (Fig. 1). For transfection of the myxoid liposarcoma cells, the calcium phosphate DNA precipitation method (Graham and van der Eb 1973) with minor modifications (Kazmierczak et al. 1990) was used.

Two foci of rapidly proliferating cells arose about 4 weeks after transfection of the primary myxoid liposarcoma cells. Cells of the foci were collected and propagated further as cell line LIS-3/SV40. Cells were passaged once or twice a week. Immunocytochemical analysis using FITC-labelled monoclonal antibodies with specificity for the SV40 large T-antigen revealed that the

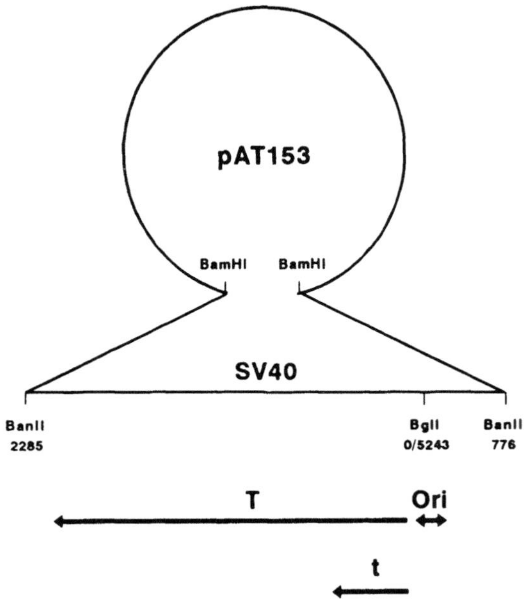

Fig. 1. Schematic representation of the organization of pSV40-ΔN DNA. For the construction of pSV40-ΔN, SV40 DNA fragment from nucleotide 776 to nucleotide 2285 (*Ban*II fragment; late region) was deleted from the viral genomic DNA, and the remaining portion of the genome (early region) was cloned into the *Bam*HI site of plasmid pAT153. The relative positions of the replication origin of SV40 and the coding sequences for the large and small T antigen are indicated

LIS-3/SV40 cells expressed large T. The LIS-3/SV40 cells appeared to possess a highly increased or even unlimited growth potential with a high proliferative capacity. This in contrast to nontransfected myxoid liposarcoma cells, the in vitro life span of which was apparently limited to only two passages. The LIS-3/SV40 cells also had a higher proliferative capacity as the nontransfected cells from the same myxoid liposarcoma, which were cultured in parallel. In Fig. 2 the karyotype of the LIS-3/SV40 cells (passage number 5) is presented. All cells appeared to contain translocation t(12;16)(q13;p11.2) and trisomy 8. Analysis of metaphases of LIS-3/SV40 cells from later passages, up to passage 116, always revealed the same karyotype, indicating that the karyotype of these cells is very stable. The observed chromosome stability makes the LIS-3/SV40 cell line suitable for use in molecular studies.

Fig. 2. Representative G-banded karyotype of LIS-3/SV40 cells, which shows the derivative chromosomes der(12) and der(16) resulting from the characteristic t(12;16)(q13;p11) of myxoid liposarcoma. LIS-3/SV40 is also characterized by a trisomy 8. *Arrows*, derivative chromosomes der(12) and der(16)

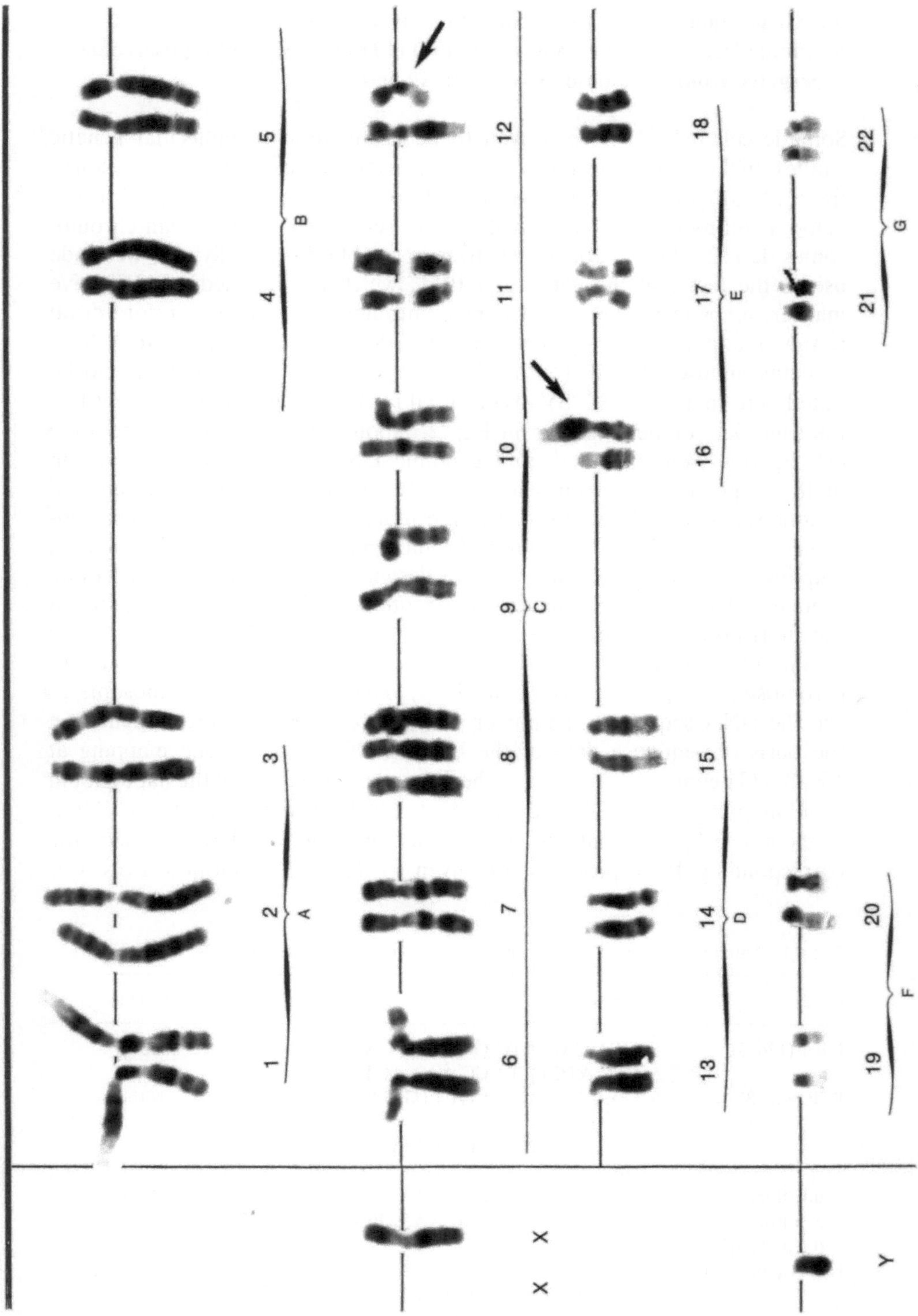
1
2
3
A
4
5
B
6
7
8
9
10
11
12
C
13
14
15
D
16
17
18
E
19
20
F
21
22
G
X X
X Y

3.2 Generation of Man/Mouse Somatic Cell Hybrids Retaining Derivative Chromosome der(16) of Human Myxoid Liposarcoma Segregated from Chromosomes der(12), 12 and 16

Somatic cell hybrids have proven to be useful tools in molecular genetic studies. In the context of studies pertaining to the molecular characterization of the t(12;16)(q13;p11) of human myxoid liposarcomas, we attempted to develop a cell hybrid that contained der(16) segregated from human chromosomes der(12), 12 and 16. In the strategy to obtain such a hybrid, we made use of the fact that the gene for APRT, which can be used as a selective marker, maps in region q24 of human chromosome 16 and therefore distal to the breakpoint in chromosome 16. LIS-3/SV40 cells were fused to APRT-defective mouse A9 cells (outlined in more detail in Sect. 2). It should be noted here that the LIS-3/SV40 cells used for the cell fusion experiment had the same karyotype as shown in Fig. 2. About 60 colonies of proliferating cell hybrids were obtained, and genomic DNA of these was assayed in multiplex polymerase chain reaction (PCR) experiments for the presence of genetic sequences known to map in the **p** and **q** regions of human chromosomes 12 and 16. To detect human chromosome 12 specific DNA sequences, amplimer sets were used which were designed on the basis of nucleotide sequence data of the pregnancy zone protein (PZP) gene, which maps on the short arm of chromosome 12 (12p13) (Zhang et al. 1990), and the gene insulin-like growth factor 1 (IGF-1) gene, which maps on the long arm of chromosome 12 (12q23) (Dull et al. 1984). To detect human chromosome 16 specific DNA sequences, amplimer sets were used which were designed on the basis of sequence data of the haemoglobin alpha 1 gene mapping at 16p13.3 (Theune et al. 1991), and the 3′ non-coding region of the haptoglobin gene mapping at 16q22.1 (Bensi et al. 1985; Abbott and Povey 1991). Nucleotide sequence data of the various amplimers and the sizes of the corresponding PCR products are given in Table 1. Multiplex PCR was

Table 1. Amplimer sets used in PCR characterization of somatic cell hybrids

Locus	Amplimer set	Product length
IGF1 (12q23)	5′ GCA CAA ATG GAT TAT CAA GG 3′ 5′ CTC ATC CTT CCC AAA TGT CA 3′	552 bp
PZP (12p13)	5′ GAA TTC ACC ATA TTG GAA GCA GGA CGG TTT 3′ 5′ CGT TCA GGA TTC ACA GAC CTT TTC TAC AAA TT 3′	990 bp
Haptoglobin 3′ untranslated region (16q22.1)	5′ GAA CTA ATG CAA GGC TGG CCG GAA 3′ 5′ GAT TGA CTC AGC AAT GCA GGG CTG G 3′	148 bp
HBA1 (16p13.3)	5′ ATC AAC TGT CAG GAA GAC GGT GTC 3′ 5′ TAC AGG CAT GAG TCA TCA CAC CTG 3′	350 bp

performed according to standard conditions; reaction mixtures (100 μl) contained 5 mM KCl, 10 mM Tris-HCl (pH 8.3), 1.5 mM $MgCl_2$, 0.01% gelatine, 0.2 mM of each of the dNTPs, 20 pmol of each primer, 2.5 U Amplitaq (Perkin Elmer, Cetus) and approximately 500 ng genomic DNA isolated from the various man/mouse cell hybrids; the amplification profile was as follows: initial denaturation for 5 min at 94°C, 30 cycles of 1 min at 94°C, 1 min at 62°C, and 1 min at 68°C, and final extension for 5 min at 68°C. In Fig. 3, results of PCR analysis of one of the selected clones (cell hybrid LIS-3/SV40/A9-B4) and appropriate controls (cell lines A9 and LIS-3/SV40) are shown as an illustration. By the PCR analysis four colonies were identified that were positive for IGF-1 and the non-coding sequences of haptoglobin but negative for PZP and the haemoglobin alpha 1 genetic sequences; these cell hybrids were designated LIS-3/SV40/A9-B4, LIS-3/SV40/A9-D3, LIS-3/SV40/A9-F4, and LIS-3/SV40/A9-H2, and they were propagated independently. The PCR results suggested that these four hybrid cell lines contained derivative chromosome der(16) but not the chromosomes der(12), 12 and 16. Karyotyping of G-banded metaphases of the cell hybrids was performed to establish the number and nature of the human chromosomes present; an example of the karyotype of a cell of cell hybrid LIS-3/SV40/A9-B4 is shown in Fig. 4. The cytogenetic studies revealed that derivative chromosome der(16) was present in all cells of each of the four cell hybrids, and that a number of human chromosomes, including der(12), 12 and 16 were missing. Screening of 20 metaphases of LIS-3/SV40/A9-B4 cells that were cultured for a prolonged period of time under selective conditions (AOA medium) indicated that in addition to the loss of human chromosomes 12, 16 and der(12) also chromosomes 9, 13, 14, 22 and Y were lost (Fig. 5).

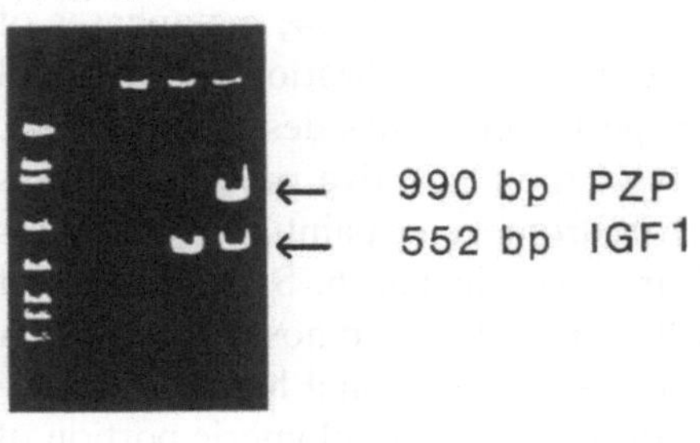

Fig. 3. Multiplex PCR analysis of genomic DNA isolated from mouse cell line A9 (*lane 3*), man/mouse somatic cell hybrid LIS-3/SV40/A9-B4 (*lane 4*), or the human myxoid liposarcoma cell line LIS-3/SV40 (*lane 5*). PCR analysis was performed as described in Sect. 2 using the amplimer sets for PZP and IGF-1. *Lane 1*, pGEM molecular weight markers 2645, 1065, 1198, 676, 517, 460, 396, and 350 bp; *lane 2*, negative control, no genomic DNA added. *Arrows*, the size of two PCR products; 990 bp with PZP amplimers, 552 bp with IGF-1 amplimers

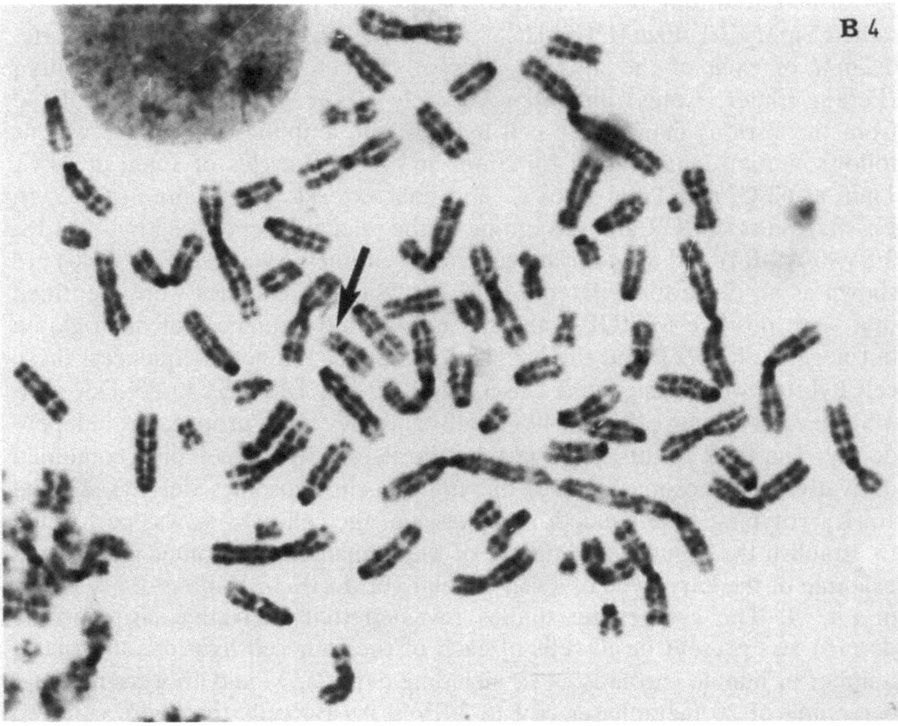

Fig. 4. Illustrative G-banded karyoptype of a metaphase of the somatic cell hybrid LIS-3/SV40/A9-B4. *Arrow*, derivative chromosome der(16)

3.3 Characterization of LIS-3/SV40 and LIS-3/SV40/A9-B4 by Chromosome Painting

To further characterize LIS-3/SV40 and LIS-3/SV40/A9-B4, metaphases of these cells were analysed by fluorescence in situ hybridization using a pool of chromosome 12 and chromosome 16 specific probes as described in Sect. 2. In control experiments, the specificity of the respective probe pools was established (data not shown). The results of chromosome painting experiments of LIS-3/SV40 and LIS-3/SV40/A9-B4 are given in Fig. 6. Such analysis of metaphase chromosomes of LIS-3/SV40 using the chromosome 12 DNA pool as probe revealed strong staining of the entire normal human chromosome 12, a large part of der(12) except for a substantial telomeric portion of the **q** arm and a telomeric portion of the **p** arm of der(16). It should be noted that some background staining is visible on all chromosomes. Similar analysis with the pool of cosmids from chromosome 16 revealed strong staining of the entire chromosome 16, a large portion of der(16) except for a telomeric part of the **p** arm and a telomeric part of the **q** arm of der(12).

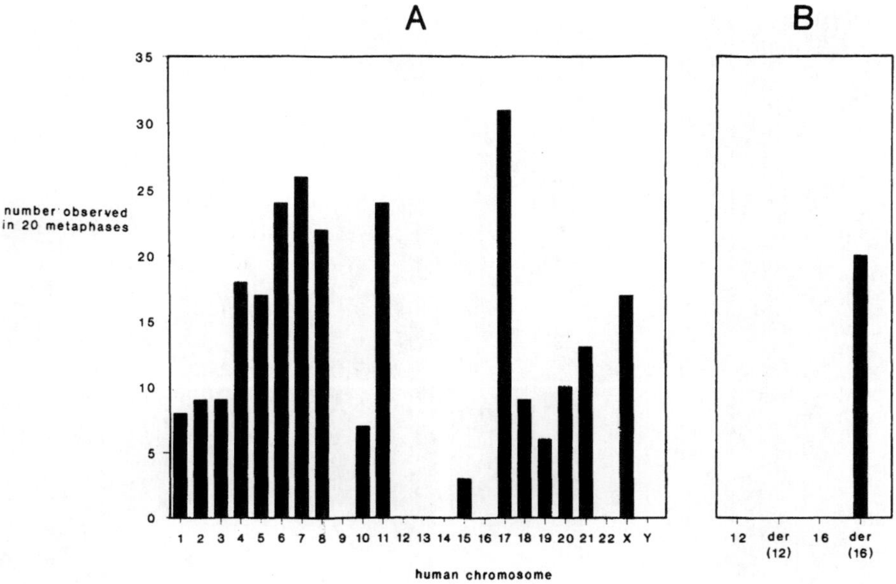

Fig. 5. A Presence of human chromosomes in somatic cell hybrid LIS-3/SV40/A9-B4 as deduced from data obtained from the analysis of 20 metaphase spreads. **B** Of special importance here is the presence of der(16) and the loss of derivative chromosome der(12) and the normal chromosomes 12 and 16

Some background staining of all chromosomes is also visible in these experiments. Chromosome painting analysis of metaphases of somatic cell hybrid LIS-3/SV40/A9-B4 with the same molecular probes revealed only strong painting of derivative chromosome der(16). This result confirms the cytogenetic observation that chromosomes der(12), 12 and 16 were not present. Altogether, the studies of LIS-3/SV40/A9-B4 suggest that most if not all of its chromosome 12 and chromosome 16 specific genetic sequences are located in derivative chromosome der(16).

4 Concluding Remarks

This study describes the development of the LIS-3/SV40 cell line from a primary myxoid liposarcoma. Using a similar approach as in earlier studies, cell lines were established from primary benign solid tumours such as pleomorphic adenoma of the salivary gland (Kazmierczak et al. 1990), uterine leiomyoma (Stern et al. 1991) and lipoma (Belge et al. 1992). During the process of establishing cell lines from the three primary benign tumours and the primary myxoid liposarcoma, one remarkable difference was noted. Upon transfection of primary tumour cells of the three benign tumours, the

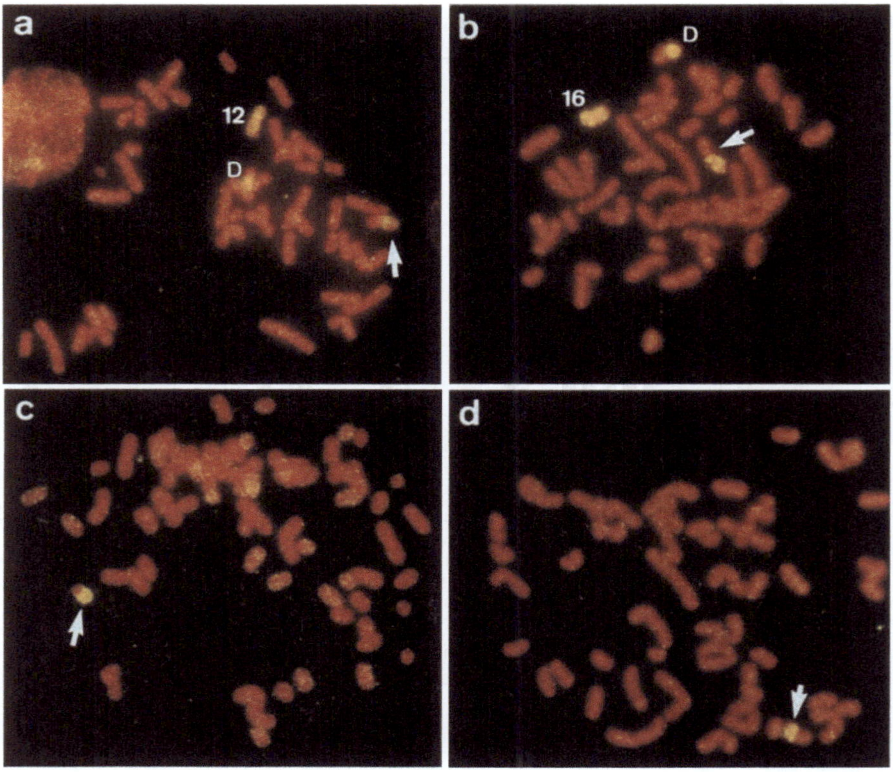

Fig. 6. Chromosome painting analysis of cell line LIS-3/SV40 (**a,b**) and somatic cell hybrid LIS-3/SV40/A9-B4 (**c,d**) using pBS12 DNA (**a,c**) or DNA of a pool of 166 cosmid clones from human chromosome 16 (**b,d**) as molecular probes. *12*, *16*, Chromosomes 12 and 16 are indicated by their number; *D*, derivative chromosome der(12); *arrows*, derivative chromosome der(16)

majority of the cells entered in a crisis which was characterized by a highly reduced proliferative capacity; after this period of crisis, cells with a higher proliferative capacity emerged; these cells showed a drastically increased or even unlimited growth potential in vitro and formed colonies in soft agar. In contrast, the transfected primary myxoid liposarcoma cells did not go through a crisis period like the cells derived from the three benign tumours. This variation in response to transfection of the SV40 "early region" suggests a direct cooperative effect of the viral genetic sequences in myxoid liposarcoma cells whereas in the benign tumours an additional genetic event seems to be necessary. The observed difference in response might also be indicative for differences in tumour development.

Of special interest is the observation that the karyotype of the LIS-3/SV40 cell line is very stable; cytogenetic analysis of cells with passage number up to 116 appeared to have the same karyotype as the primary tumour cells. The stability of the karyotype during a prolonged period of time is important and makes the cells useful for the construction of genomic and cDNA libraries, experiments for which relatively large numbers of cells are required.

The LIS-3/SV40 and the LIS-3/SV40/A9-B4 cell lines that have become available through the studies described here constitute valuable reagents for the molecular analysis of the chromosome breakpoint t(12;16)(q13;p11) of myxoid liposarcoma. First of all, as already shown above, genomic DNA isolated from LIS-3/SV40/A9-B4 cells can be used quickly to map genetic sequences with respect to the breakpoint; this can be achieved by PCR or Southern blot analysis. Furthermore, the high mitotic index of both cell lines makes the cells especially suitable for mapping newly isolated DNA probes relative to the translocation breakpoint of myxoid liposarcoma by fluorescence in situ hybridization analysis. Finally, the availability of the cell lines also facilitates Southern blot analysis of the DNA sequences that are directly affected by the translocations and, eventually, the molecular cloning and characterization of the chromosome breakpoints in chromosome 12 and 16.

Acknowledgements. The authors wish to thank A.A. Sandberg for providing the primary myxoid liposarcoma tumour sample, J.W. Gray (UCSF) for providing pBS12, P. Devilee (Leiden, The Netherlands) for the collection of chromosome 16 cosmids, M. Willems for their contribution in generating the somatic cell hybrids, and S. Bartnitzke and M. Dehaen for excellent technical assistance in the cytogenetic studies. This work was supported in part by the EC Medical and Health Research Programme, the EC-sponsored concerted action program "Molecular Cytogenetics of Solid Tumours", by the Inter-University Network for Fundamental Research sponsored by the Belgian Government (1991–1995), and the "Geconcerteerde Onderzoekacties 1992–1996", and the Schwerpunktprogramm "Molekulare und Klassische Tumorcytogenetik" of the Deutsche Forschungsgemeinschaft.

References

Abbott C, Povey S (1991) Development of human chromosome-specific PCR primers for characterization of somatic cell hybrids. Genomics 9:73–77

Belge G, Kazmierczak B, Miyir-Bolte K, Bartnitzke S, Bullerdiek J (1992) Expression of SV40 T antigen in lipoma cells with the chromosomal translocation t(3;12) is not sufficient for direct immortalization. Cell Biol (Int Rep) 16:4

Bensi G, Raugei G, Klefenz H, Cortese R (1985) Structure and expression of the human haptoglobin locus. EMBO J 4:119–126

Collins C, Kuo WL, Segraves R, Pinkel D, Fuscoe J, Gray JW (1991) Construction and characterization of plasmid libraries enriched in sequences from single human chromosomes. Genomics 11:997–1006

De la Maza M, Sanchez O (1976) Simultaneous G and C banding of human chromosomes. J Med Genet 13:235–236

Dull TJ, Gray A, Hayflick JS, Ullrich A (1984) Insulin-like growth factor II precursor gene organization in relation to insulin gene family. Nature 310:770–781

Graham FL, van der Eb AJ (1973) A new technique for the assay of infectivity of human adenovirus 5 DNA. Virology 52:456–467

Kazmierczak B, Bartnitzke S, Hartl M, Bullerdiek J (1990) In vitro transformation by the SV40 "early region" of cells from a human benign salivary gland tumour with a 12q13→q15 rearrangement. Cytogenet Cell Genet 53:37–39

Kievits T, Dauwerse JG, Wiegant J, Devilee P, Breuning MH, Cornelisse CJ, van Ommen G, Pearson PL (1990) Rapid subchromosomal localization of cosmids by nonradioactive in situ hybridization. Cytogenet Cell Genet 53:134–136

Mitelman F (1991) Catalog of chromosome aberrations in cancer, 3rd edn. Liss, New York

Ohjimi Y, Iwasaki H, Kaneko Y, Ishiguro M, Ohgami A, Fujita C, Shinohara N, Yoshitake K, Kikuchi M (1992) Chromosome abnormalities in liposarcomas. Cancer Genet Cytogenet (in press)

Pandis N, Heim S, Willen H, Bardi G, Floderus UM, Mandahl N, Mitelman F (1991) Chromosome analysis of 96 uterine leiomyomas. Cancer Genet Cytogenet 55:11–18

Pontecorvo G (1975) Production of mammalian somatic cell hybrids by means of polyethylene glycol treatment. Somat Cell Genet 1:397–400

Sandberg AA (1990) The chromosome in human cancer and leukemia. Elsevier, New York

Sandros J, Stenman G, Mark J (1990) Cytogenetic and molecular observations in human and experimental salivary gland tumours. Cancer Genet Cytogenet 44:153–167

Solomon E, Borrow J, Goddard AD (1991) Chromosome aberrations and cancer. Science 254:1153–1160

Sreekantaiah C, Leong SLP, Karakousis CP, McGee DL, Rappaport WD, Villar HV, Neal D, Fleming S, Wankel A, Herrington PN, Carmona R, Sandberg AA (1991) Cytogenetic profile of 109 lipomas. Cancer Res 51:422–433

Stern C, Kazmierczak B, Thode B, Rommel B, Bartnitzke S, Dal Cin P, Van de Ven W, Van den Berghe H, Bullerdiek J (1991) Leiomyoma cells with 12q15 aberrations can be transformed in vitro and show a relatively stable karyotype during precrisis period. Cancer Genet Cytogenet 54:223–228

Theune S, Fung J, Todd S, Sakaguchi AY, Naylor SL (1991) PCR primers for human chromosomes: reagents for the rapid analysis of somatic cell hybrids. Genomics 9:511–516

Turc-Carel C, Limon J, Dal Cin P, Rao U, Karakousis C, Sandberg AA (1986a) Cytogenetic studies of adipose tissue tumours. II. Recurrent reciprocal translocation t(12;16)(q13;p11) in myxoid liposarcomas. Cancer Genet Cytogenet 23:291–299

Turc-Carel C, Dal Cin P, Rao U, Karakousis C, Sandberg AA (1986b) Cytogenetic studies of adipose tissue tumours. I. A benign lipoma with reciprocal translocation t(3;12)(q28;q14). Cancer Genet Cytogenet 23:283–289

Zhang J, Devriendt K, Marynen P, Van den Berghe H, Cassiman JJ (1990) Chromosome mapping using polymerase chain reaction on somatic cell hybrids. Cancer Genet Cytogenet 45:217–221

Amplification of the *GLI* and *LRP/A2MR* Loci in Tumor Cells: Is *GLI* only by Chance Coamplified Together with Another Gene Related to Tumor Progression?

A. Forus[1], G.M. Maelandsmo[1], Ø. Fodstad[1], E.C. Douglass[2], and O. Myklebost[1]

1 Introduction

One property of tumor cells is their genomic instability. This is observed at several levels, as increased mutation and amplification frequency, chromosomal deletions, duplications, and translocations. We are particularly interested in translocations and amplifications of the q13–14 segment of chromosome 12. Aberrations of the segment q13–15 on chromosome 12 has been found in several tumor types, notably benign lipomas and (malignant) myxoid liposarcomas, as well as in uterine leoimyomas and pleomorphic adenomas of the salivary gland (Turc-Carel et al. 1986; Bullerdiek et al. 1987; Mandahl et al. 1987; Heim et al. 1988). The most specific of these rearrangements are those found in the myxoid liposarcomas, where the distal part of chromosome 12 (q13-qter) is reciprocally translocated to 16p11. Several putative oncogenes have been localized to this chromosomal region and thus could be candidate target genes for the rearrangements.

The *GLI* gene, situated in this region (Arheden et al. 1989) and coding for a zinc finger protein, was initially cloned because of its amplification in a glioma (Kinzler et al. 1987; Ruppert et al. 1988). Although amplification of *GLI* seems to be a rare event, it has been found to be amplified also in some childhood sarcomas (Roberts et al. 1989). When transfected, *GLI* can transform primary cells in cooperation with the adenovirus E1A gene (Ruppert et al. 1991).

Another interesting gene found in this region is that coding for the low-density lipoprotein (LDL) receptor-related protein (LRP; Myklebost et al. 1989). LRP has structural homologies both to the LDL receptor and to epidermal growth factor (EGF) and its precursor molecule (Herz et al.

[1] Department of Tumor Biology, Institute for Cancer Research, Norwegian Radium Hospital, Montebello, N-0310 Oslo, Norway

[2] Department of Hematology/Oncology, St. Jude Children's Research Hospital, Memphis, Tennessee, USA

1988). There is evidence for LRP being an apolipoprotein E receptor (Beisiegel et al. 1989; Kowal et al. 1989), but its physiological role in lipid metabolism is at present unclear. LRP is cleaved in the endosomal compartment, resulting in an 80-kDa membrane-spanning peptide still associated with the 500-kDa peptide containing the ligand-binding domains (Lund et al. 1989; Herz et al. 1990).

Surprisingly, two independent groups have recently shown that LRP is identical to the α_2-macroglobulin (α2M) receptor (Kristensen et al. 1990; Strickland et al. 1990). This receptor (α2Mr) is responsible for the internalization and subsequent degradation of the protease inhibitor α2M complexed to inactivated proteases (see Sottrup-Jensen 1989 for review). It is therefore likely that α2Mr/LRP is bifunctional in a similar way as is the mannose-6-phosphate receptor (Tong et al. 1988; MacDonald et al. 1988) although it might be difficult to envisage how the two processes would be coordinated. However, the complex structure of LRP, consisting of 4544 amino acids plus an additional associated chain (Herz et al. 1991) and with several putative ligand-binding domains, would be a good candidate for a receptor with more than one function. Interestingly, just as the LDL receptor and one of its ligands, apoE, are localized on the same chromosome (Francke et al. 1984), the α2Mr and α2M are also localized to one chromosome (Bell et al. 1985; Myklebost et al. 1989). Because the role of LRP/α2Mr in lipoprotein metabolism is at present unclear, whereas the protein clearly functions as a receptor for α2M, it is more relevant to call the locus A2MR rather than its original HGM symbol, "apolipoprotein receptor" (APR).

α2M is reported to bind growth factors, and it is therefore possible that the activity of α2Mr could modulate growth (see Gonias 1992 for review). Furthermore, α2M may be involved in the metastatic process (see Kwaan 1988 for review). Although it is conceivable that disturbances of the function of an apoE receptor could be partly responsible for the adipocytelike morphology of lipoma and liposarcoma cells, aberrant expression of LRP in such tumors has not been observed (GM and OM, unpublished observations). It thus is more likely that the translocations in lipogenic tumors affect other as yet uncharacterized genes in the same chromosomal region which are important for lipid metabolism or adipocyte differentiation.

2 Results and Discussion

2.1 Long-Range Restriction Mapping

By pulsed field gradient (PFG) gel analysis (Fig. 1) we have found that the *GLI* and *LRP* genes are located close together. Probes for either gene hybridize to DNA fragments of molecular weight of 340 kb generated by

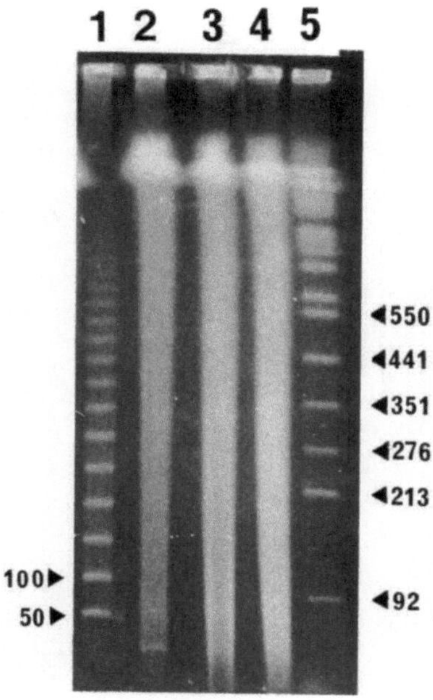

Fig. 1. Ethidium bromide fluorescence from a typical PFG gel. Leukocytes were lysed within agarose blocks and protein removed by treatment with protease K and extensive washing with detergent and EDTA (Myklebost and Rogne 1988). Agarose-embedded DNA samples (Lane 2 to 4) were digested with rare-cutting restriction enzymes and separated on a PFG gel with pulse times ramped from 5 to 90 s. *Lane 1*, concatemers of bacteriophage λDNA, each band is a multiple of 50 kb; *Lane 5*, yeast chromosomes from *Saccharomyces cerevisiae* strain YPH148, molecular weights range from 92 kb to more than one megabase

cutting DNA from leukocytes with either *Csp*45I or *Sfu*I (Forus et al. 1991). Partial digestion patterns confirm that the fragments hybridizing to both probes are identical (Fig. 2a). Furthermore, in DNA from sperm both probes hybridize to a *Csp*45I fragment of about 500 kb (Fig. 2b), probably due to different methylation of cytosines in the recognition sequence. This genomic region is also quite variably methylated in tumor cells. In some tumor samples the fragments hybridizing to LRP and GLI probes are very small, probably due to demethylation. However, most tumors that we have examined have *Csp*45I fragments containing *GLI* and *LRP* which are larger than those in leukocytes, and the identical hybridization pattern obtained for the two probes also in these tumors confirm that the probes hybridize to the same *Csp*45I fragment (Fig. 3). A more detailed restriction analysis (Forus and Myklebost 1992) shows that the intergenic region is between 200 and 300 kb. Between the two genes there is a metylation-free island, commonly associated with active genes (Bird 1986).

2.2 Amplification Analysis

Amplification of *GLI* has been described for several childhood sarcomas, for example, the osteosarcoma cell line OSA and the rhabdomyosarcoma cell line RMS13 (Roberts et al. 1989; Fig. 4a). In the RMS13 cell line also the LRP/A2MR-gene was amplified (Fig. 4b). We have analyzed the copy number of the two genes by laser densitometry of the autoradiograms. The blots were also hybridized to a control probe, and the integrated signal from

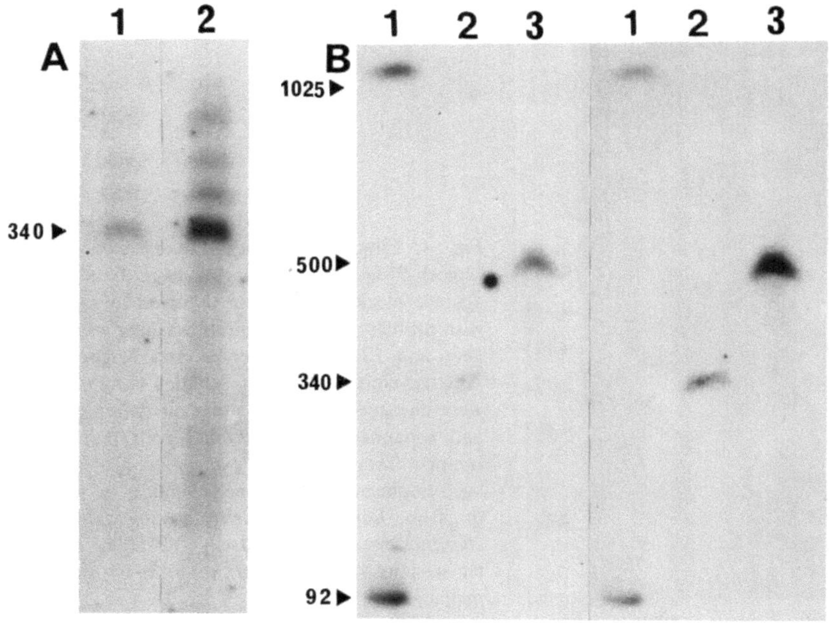

Fig. 2A,B. Autoradiograms of PFG gels with partially cut DNA and DNA from sperm cells hybridized sequentially to probes for the *GLI* and *LRP* genes. **A** Incomplete digestion with *Csp*45I (recognition sequence TTCGAA, = *Sfu*I) shows identical patterns of partially cut fragments above the completely digested 340-kb band for both GLI (*lane 1*) and LRP (*lane 2*). **B** Both the GLI and LRP probes hybridize to a *Csp*45I fragment of 500 kb in sperm DNA (*lane 3*). The altered fragment size is probably caused by differential methylation in sperm cells. *Lane 2* shows the 340 kb fragment in leukocyte DNA from the same person, and *lane 1* contains yeast chromosomes

GLI- and *LRP*-specific bands were corrected relative to the signals obtained from the control probe. *GLI* was amplified about 20-fold in the OSA line (Fig. 5), whereas the *GLI* gene is amplified about 15-fold in RMS13; the amplification of the *LRP* gene is about half of this.

Due to the saturation of the autoradiographic film at the high optical densities obtained from the amplified samples, there is some uncertainty about the measured degrees of amplification. However, control experiments, where known amounts of radioactive DNA was bound to filters in dot blots, indicate that when scanning is done in this way, there is an almost linear relationship between the values obtained and the actual radioactivity on the filter. This is because the total signal from the band increases as the band widens, although the peak optical density in the center of the band is becoming saturated. When identical areas large enough to include the widest bands are measured for the bands in amplified and control samples, the net signals after background subtraction should give a reasonable measure of the relative copy numbers.

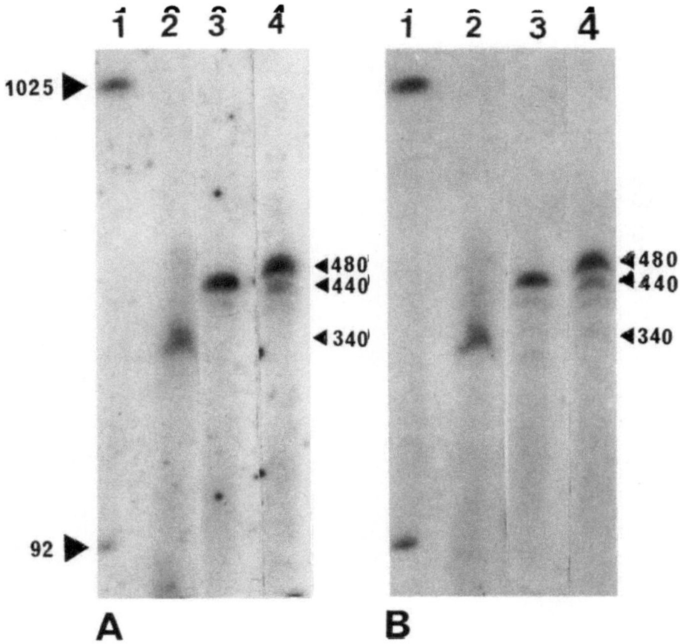

Fig. 3A,B. Autoradiograms of a PFG gel with DNA from various tumor cells, hybridized sequentially to probes for the *GLI* and *LRP* genes. Digestion with *Csp*45I shows identical fragments for both *GLI* (**A**) and *LRP* (**B**). Both the GLI and LRP probes hybridize to *Csp*45I fragments of 340 kb in leukocytes (*lane 2*), 440 kb in DNA from a breast cancer cell line (*lane 3*), and 480 kb (plus some smaller bands) in DNA from a melanoma cell line (*lane 4*). The altered fragment size is probably caused by differential methylation. *Lane 1* contains yest chromosomes

The difference in copy number observed for these two genes within the same amplification unit is consistent with detailed studies of the degree of amplification of different sequences along other amplification units which were experimentally induced (Debatisse et al. 1986; Saito et al. 1989). In these studies, during the first rounds of amplifications large segments of DNA (typically several hundred kilobases) were amplified. After a long time under selection the complexity of the amplicon was reduced, and it contained sequences flanking the selected gene in higher copy numbers. If there is a target gene near *LRP* or *GLI* offering some selective advantage when amplified, one might speculate that it is closer to *GLI* than to *LRP*.

2.3 Characterization of the Minimal Amplicon

We have found a liposarcoma tumor with amplification of *LRP* but not *GLI*. PFG mapping of this region in that tumor revealed that only the 3′

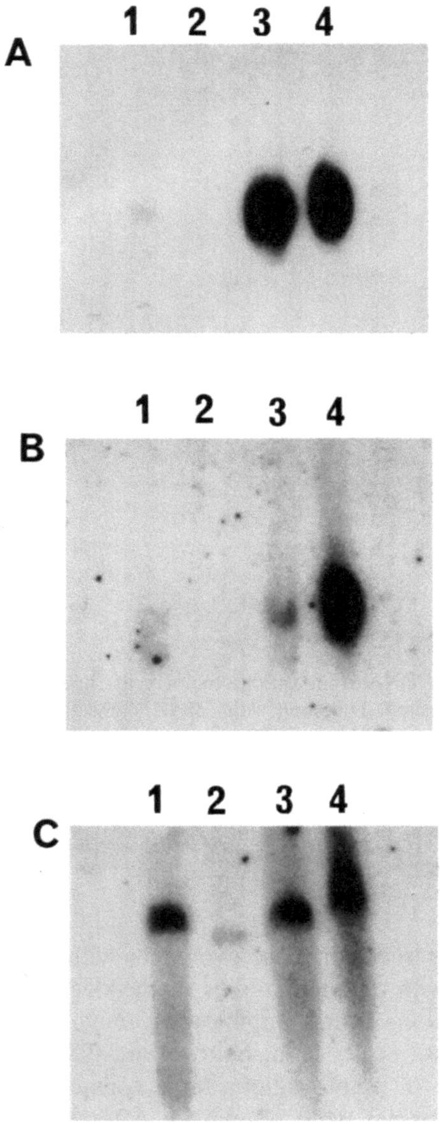

Fig. 4 A–C. Autoradiograms of a PFG gel containing samples with amplifications of the chromosomal region containing the *GLI* and *LRP* genes. The same gel was sequentially hybridized to probes for GLI (**A**), LRP (**B**), and an unamplified gene (7.16.1, **C**). Samples were normal leukocyte DNA (*lane 1*), a lipoma (*lane 2*), osteosarcoma OSA (*lane 3*), and rhabdomyosarcoma RMS13 (*lane 4*). Quantitative results from this gel are shown in Fig. 5

end of *LRP* is amplified. Combining the results from OSA, RMS-13, and this liposarcoma localizes the minimal overlap of the amplified regions to the area between *LRP* and *GLI*. We are now searching for new genes in this region and hope to find one more consistently amplified in sarcomas.

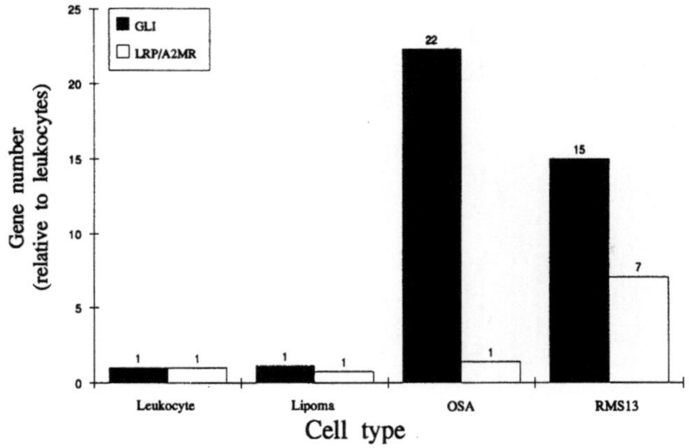

Fig. 5. Densitometric analysis of degree of amplification of the *GLI* and *LRP* genes. The autoradiograms shown in Fig. 4 were scanned with a laser densitometer. The optical density of individual bands (after background subtraction) was corrected for variations in amount of DNA in the lane as measured by hybridization to the probe for the unamplified gene (Fig. 4C)

2.4 Expression Analysis

Compared to other rhabdomyosarcoma cell lines LRP/α2Mr mRNA levels were increased proportionally to the degree of amplification in the RMS13 cells, whereas GLI expression was increased only 4- to 5-fold both in OSA and RMS13 cells (Fig. 6, 7). A xenograft (Rh30) established from the same tumor as the RMS13 cell line had lower levels of expression of both the *GLI* and *LRP/A2MR* genes. At present it is unknown whether the degree of amplification is the same in the xenograft as in the corresponding cell line. The level of amplification in the cell line is similar to that found for the original tumor, and probably the differences found between the xenograft and the cell line reflects differences in growth conditions in vivo and in vitro.

3 Conclusions

At present it is not known whether the amplified region includes even more genes, or whether the rearrangements or increased expression levels of *LRP* or *GLI* significantly affect the biological properties of the tumor cells. Because the GLI gene has oncogenic potential and some cells, such as the OSA cell line, has amplified *GLI* but not *LRP/A2MR*, it seems more likely that the amplification of *GLI* is important for the oncogenic process. However, it is quite likely that these two genes are only occasionally coamplified together with an unidentified gene close by. This could explain the low frequency of tumors with *GLI* amplification and our finding in a liposarcoma of amplification of *LRP* but not *GLI*.

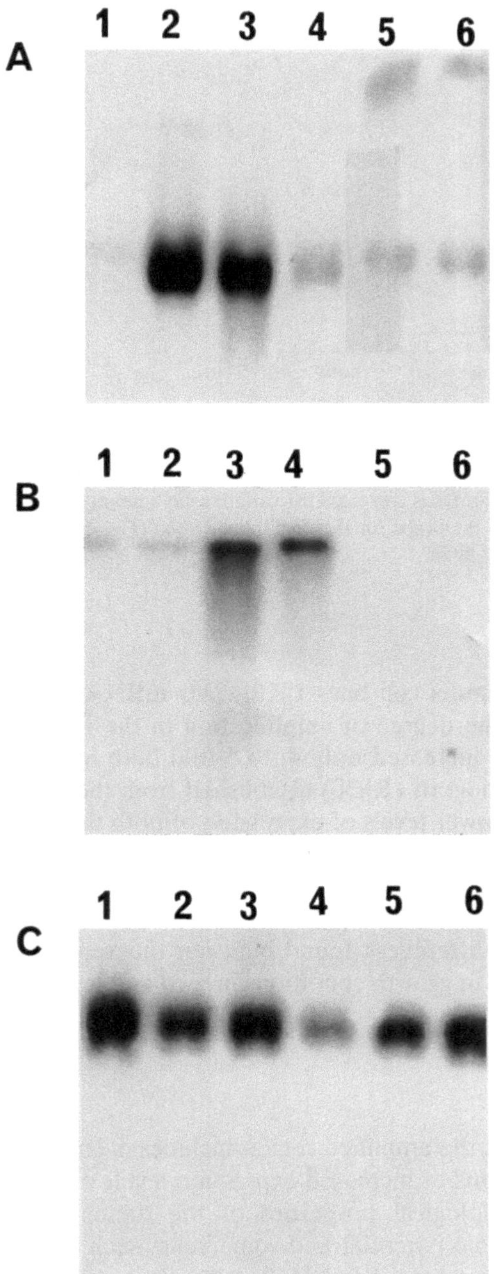

Fig. 6A–C. Autoradiograms of blot containing RNA from tumor cell lines with amplification of the *GLI* or *LRP* genes. The blot was sequentially hybridized with probes for GLI (**A**), LRP/A2MR (**B**), and 18S ribosomal RNA (for calibration, **C**). Samples were leukocytes (*lane 1*), osteosarcoma OSA (*lane 2*), rhabdomyosarcoma RMS13 (*lane 3*), rhabdomyosarcoma xenograft Rh30 (*lane 4*), rhabdomyosarcomas Rh28 and Rd/a (*lanes 5, 6*). Quantitations from these blots are shown in Fig. 7

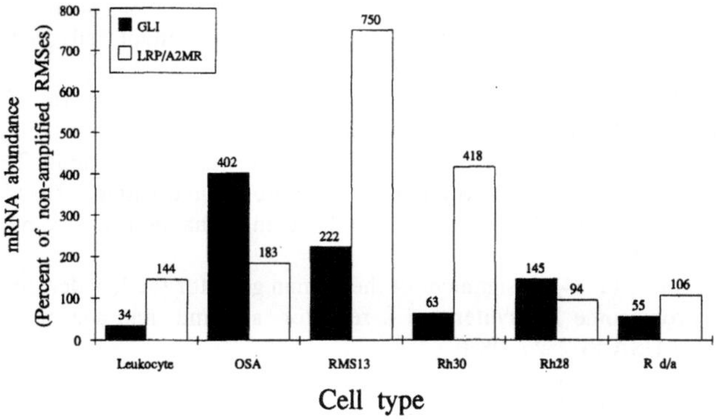

Fig. 7. Densitometric analysis of the expression levels of the GLI and LRP/A2MR genes. The autoradiograms shown in Fig. 6 were scanned with a laser densitometer. The optical density of individual bands (after background subtraction) was corrected for variations in amount of RNA in the lane as measured by hybridization to a probe for ribosomal RNA (Fig. 6C). The mRNA levels are expressed as percent of the average level found in the rhabdomyosarcomas without amplification of the two genes

Another gene in the 12q13–14 region which is frequently amplified in sarcomas is the SAS gene (Meltzer et al. 1991). It has not yet been possible to analyze our samples for amplification of SAS, but this gene is probaly not close to *GLI*, as it was not coamplified with *GLI* in any of the tumors tested (Meltzer et al. 1991). Another possibility is that the amplifications observed in this region are unspecific aberrations of tumor cells, and that they have only minor effects on the phenotypes of the cancer cells.

Acknowledgements. This work was supported by the Norwegian Cancer Research Fund, the Norwegian Research Council for Science and the Humanities and the University of Oslo.

References

Arheden K, Rönne M, Mandahl N, Heim S, Kinzler KW, Vogelstein B, Mitelman F (1989) In situ hybridization localizes the human putative oncogene GLI to chromosome subbands 12q13.3–14.1. Hum Genet 82:1–2

Beisiegel U, Weber W, Ihrke G, Herz J, Stanley KK (1989) The LDL-receptor-related protein, LRP, is an apolipoprotein E-binding protein. Nature 341:162–164

Bell GI, Rall LB, Sanchez-Pescador R, Merry-Wheather JP, Scott J, Eddy RL, Shows TB (1985) Human alpha2-macroglobulin gene is located on chromosome 12. Somat Cell Mol Genet 11:285–289

Bird AP (1986) CpG-rich islands and the function of DNA methylation. Nature 321:209–213

Bullerdiek J, Bartnitzke S, Weinberg M, Chilla R, Haubrich J, Schloot W (1987) Rearrangements of chromosome region 12q13–15 in pleomorphic adenomas of the salivary gland (PAS). Cytogenet Cell Genet 45:187–190

Debatisse M, Hyrien O, Petit-Koskas E, Robert de Saint-Vincent B, Buttin G (1986) Segregation and rearrangement of coamplified genes in different lineages of mutant cells that overproduce adenylate deaminase. Mol Cell Biol 6:1776–1781

Forus A, Myklebost O (1992) A physical map of a 1.3 megabase region on the long arm of chromosome 12, spanning the GLI and LRP loci. Genomics 14:117–120

Forus A, Maelandsmo GM, Fodstad Ø, Myklebost O (1991) The genes for the alpha-2-macroglobulin receptor/LDL receptor-related protein and GLI are located within a chromosomal segment of about 300 kilobases and are coamplified in a rhabdomyosarcoma cell line. Cytogenet Cell Genet 58:1977

Francke U, Brown MS, Goldstein JL (1984) Assignment of the human gene for the low density lipoprotein receptor to chromosome 19: synteny of a receptor, a ligand, and a genetic disease. Proc Natl Acad Sci USA 81:2826–2830

Gonias SL (1992) Alpha2-macroglobulin: a protein at the interface of fibrinolysis and cellular growth regulation. Exp Hematol 20:302–311

Heim S, Nilbert M, Vanni R, Flodérus U-M, Mandahl N, Liedgren S, Lecca U, Mitelman F (1988) A specific translocation, t(12;14)(q14–15;q23–24), characterizes a subgroup of uterine leiomyomas. Cancer Genet Cytogenet 32:13–17

Herz J, Hamann U, Rogne S, Myklebost O, Gausepohl H, Stanley KK (1988) Surface location and high affinity for calcium of a 500 kDa liver membrane protein closely related to the LDL-receptor suggest a physiological role as lipoprotein receptor. EMBO J 7:4119–4127

Herz J, Kowal RC, Goldstein JL, Brown MS (1990) Proteolytic processing of the 600 kd low density lipoprotein receptor-related protein (LRP) occurs in a trans-Golgi compartment. EMBO J 9:1769–1776

Herz J, Goldstein JL, Strickland DK, Ho YK, Brown MS (1991) 39-kDa protein modulates binding of ligands to low density lipoprotein receptor-related protein/alpha2-macroglobulin receptor. J Biol Chem 266:21232–21238

Kinzler KW, Bigner SH, Bigner DD, Trent JM, Law ML, O'Brien SJ, Wong AJ, Vogelstein B (1987) Identification of an amplified, highly expressed gene in a human glioma. Science 236:70–73

Kowal RC, Herz J, Goldstein JL, Esser V, Brown MS (1989) Low density lipoprotein receptor-related protein mediates uptake of cholesteryl esters derived from apoprotein E-enriched lipoproteins. Proc Natl Acad Sci USA 86:5810–5814

Kristensen T, Moestrup SK, Gliemann J, Bendtsen L, Sand O, Sottrup-Jensen L (1990) Evidence that the newly cloned low-density-lipoprotein receptor related protein (LRP) is the alpha-2-macroglobulin receptor. FEBS Lett 276:151-155

Kwaan HC (1988) Tumor growth and metastases in malignant disease. A review. J Med 19:179–191

Lund H, Takahashi K, Hamilton RL, Havel RJ (1989) Lipoprotein binding and endosomal itinerary of the low density lipoprotein receptor-related protein in rat liver. Proc Natl Acad Sci USA 86:9318–9322

MacDonald RG, Pfeffer SR, Coussens L, Tepper MA, Brockleband CM, Mole JE, Anderson JK, Chen E, Czech MP, Ullrich A (1988) A single receptor binds both insulin-like growth factor II and mannose-6-phosphate. Science 239:1134–1137

Mandahl N, Heim S, Johannson B, Bennet K, Mertens F, Olsson G, Rooser B, Rydholm A, Willen H, Mitelman F (1987) Lipomas have characteristic structural chromosomal rearrangements of 12q13–q14. Int J Cancer 39:685–688

Meltzer PS, Jankowski SA, Dal Cin P, Sandberg AA, Paz IB, Coccia MA (1991) Identification and cloning of a novel amplified DNA sequence in human malignant fibrous histiocytoma derived from a region of chromosome 12 frequently rearranged in soft tissue tumors. Cell Growth Differ 2:495–501

Myklebost O, Rogne S (1988) A physical map of the apolipoprotein gene cluster on human chromosome 19. Hum Genet 78:244–247

Myklebost O, Arheden K, Rogne S, Geurts Van Kessel AHM, Mandahl N, Herz J, Stanley KK, Heim S, Mitelman F (1989) The gene for the human putative apoE receptor is on chromosome 12 in the segment q13–14. Genomics 5:65–69

Roberts WM, Douglass EC, Peiper SC, Houghton PJ, Look AT (1989) Amplification of the gli gene in childhood sarcomas. Cancer Res 49:5407–5413

Ruppert JM, Kinzler KW, Wong AJ, Bigner SH, Kao F-T, Law ML, Seuanez HN, O'Brien SJ, Vogelstein B (1988) The GLI-Kruppel family of human genes. Mol Cell Biol 8:3104–3113

Ruppert JM, Vogelstein B, Kinzler KW (1991) The zinc finger protein GLI transforms primary cells in cooperation with adenovirus E1A. Mol Cell Biol 11:1724–1728

Saito I, Groves R, Giulotto E, Rolfe M, Stark GR (1989) Evolution and stability of chromosomal DNA coamplified with the CAD gene. Mol Cell Biol 9:2445–2452

Sottrup-Jensen L (1989) Alpha-macroglobulins: structure, shape, and mechanism of proteinase complex formation. J Biol Chem 264:11539–11542

Strickland DK, Ashcom JD, Williams S, Burgess WH, Migliorini M, Argraves WS (1990) Sequence identity between the alpha-2-macroglobulin receptor and low density lipoprotein receptor-related protein suggests that this molecule is a multifunctional receptor. J Biol Chem 265:17401–17404

Tong PY, Tollefsen SE, Kornfeld S (1988) The cation-independent mannose 6-phosphate receptor binds insulin-like growth factor II. J Biol Chem 263:2585–2588

Turc-Carel C, Limon J, Dal Cin P, Rao U, Karakousis C, Sandberg AA (1986) Cytogenetic studies of adipose tissue tumors. II. Recurrent reciprocal translocation t(12;16)(q13;p11) in myxoid liposarcomas. Cancer Genet Cytogenet 23:291–299

Analysis of Solid Tumours with Recurrent Breakpoints in Chromosome Region 12q13–15 by Fluorescence In Situ Hybridization Using a Microclone Library

P.F.J. Kools[1], E.F.P.M. Schoenmakers[1], J. Bullerdiek[2], U. Claussen[3], B. Horsthemke[4], H. Van den Berghe[1], and W.J.M. Van de Ven[1]

1 Introduction

Genetic aberrations involving region q13–15 of human chromosome 12 are frequently observed in solid tumours (Sandberg 1990; Mitelman 1991). In myxoid liposarcoma, for instance, the primary cytogenetic change is mostly the highly consistent reciprocal translocation t(12;16)(q13;p11) or the more complex three-way translocation also involving chromosome 1 (Turc-Carel et al. 1986a; Ohjimi et al. 1992). Cytogenetic analysis of lipomas, a subgroup of benign adipose tissue tumours, has revealed that in these tumours region q13–15 of chromosome 12 is also frequently involved in reciprocal translocations; in contrast to myxoid liposarcomas, however, various chromosomes act as translocation partner (Turc-Carel et al. 1986b; Sreekantaiah et al. 1991). The same cytogenetic region of chromosome 12 is also implicated in translocations in a subclass of cytogenetically abnormal uterine leiomyomas (Pandis et al. 1991) and pleomorphic adenomas of the salivary gland (Sandros et al. 1990). Comparison of available cytogenetic data suggests that the chromosome 12 breakpoint in myxoid liposarcoma is proximal to those in the three benign tumours; this might point towards the implication of different genetic sequences. Identification and molecular characterization of the chromosome 12 breakpoints in cells of the four tumours should shed more light on this matter. For detailed molecular studies, cell lines that are derived from the primary tumours, and that have retained the derivative chromosome(s) of interest are required. In recent studies, we have developed cell lines for each of these four tumour types; this was achieved by transfecting cells from primary tumours with translocations involving chromosome 12q13–15 with a recombinant DNA construct harbouring the "early region"

[1] Center for Human Genetics, University of Leuven, Herestraat 49, 3000 Leuven, Belgium
[2] Zentrum für Humangenetik, University of Bremen, ZHG, Leobenerstrasse, 28359 Bremen, Germany
[3] Institut für Humangenetik, University of Erlangen, Schwabachanlage 10, 91054 Erlangen, Germany
[4] Institute for Human Genetics, University Clinic Essen, Essen, Germany

of the SV40 virus genome (Kazmierczak et al. 1990; Stern et al. 1991; Schoenmakers et al., this volume). In the present study we used myxoid liposarcoma cell line LIS-3/SV40 which has a t(12;16)(q13;p11.2) and a trisomy 8 (Schoenmakers et al. this volume), lipoma cell line LI-14/SV40 which has a t(3;12)(q28;q14) (Belge et al. 1992), uterine leiomyoma cell line LM-30.1/SV40 which has a t(12;14)(q14–15;q24) (Stern et al. 1991) and cell line PA-312 which was derived from a pleomorphic adenoma of the salivary gland with a t(1;12)(p21;q14) (Kazmierczak et al. 1990). As an initial approach to molecularly define the chromosome 12 breakpoints in these tumour cell lines, a microclone library was constructed with genomic DNA from chromosome fragments microdissected from region q13–15 of human chromosome 12 and painting experiments were performed with a DNA pool of this library as probe. The results of these experiments revealed differences in the relative positions of the chromosome 12 breakpoints in the myxoid liposarcoma cell line LIS-3/SV40 and the cell lines of the three benign tumours.

2 Materials and Methods

2.1 Human Tumour Cell Lines

Human tumour cell lines used included myxoid liposarcoma cell line LIS-3/SV40 (Schoenmakers et al., this volume), lipoma cell line LI-14/SV40 (Belge et al. 1992), uterine leiomyoma cell line LM-30.1/SV40 (Stern et al. 1991), and cell line PA-312 which was derived from a pleomorphic adenoma of the salivary gland (Kazmierczak et al. 1990). Cells were grown in culture medium (TC199 medium with Earle's salts supplemented with 20% fetal bovine serum, 200 IU/ml penicillin, and 200 μg/ml streptomycin at 37°C in 5% CO_2 in air.

2.2 Chromosome Preparations

Human metaphase spreads from cultured peripheral blood lymphocytes were prepared according to standard procedures. Briefly, cells were cultured in the presence of phytohaemagglutinin. Cells were synchronized using methotrexate and 5-bromo-2-deoxyuridine. Metaphase spreads from the various cell lines were prepared from cells harvested 48 h after passaging. Colcemid (25 ng/ml) was added 2 h before harvesting. After hypotonic shock in 75 m*M* KCl, cells were fixed using methanol : acetic acid and chromosome spreads were prepared on glass slides by standard procedures.

2.3 Chromosome Microdissection, Construction of Microclone Library and Chromosome Painting

The microclone library, which was designated ML12q1315, was constructed as previously described (Lüdecke et al. 1989, 1990; Senger et al. 1990). Ligation mixture (PCR-amplified genomic DNA from microdissected chromosome fragments ligated into pT7T3-18U; Pharmacia) was used to transform MAX efficiency competent DH5α cells (BRL) according to a procedure suggested by the manufacturer. Colonies from a densely grown filter were scraped off and an aliquot of this mixture was used for inoculation of 50 ml LB medium. DNA, isolated from this 50 ml culture using a standard midiscale DNA isolation protocol including a PEG precipitation step, was labelled in a PCR reaction during which biotin-11-dUTP was incorporated. The PCR reaction mixture (100 μl) contained 5 m*M* KCl, 10 m*M* Tris-HCl (pH 8.3), 1.5 m*M* $MgCl_2$, 0.01% gelatine, 0.2 m*M* dATP, dCTP, dGTP, 0.16 m*M* dTTP (Pharmacia), 0.04 m*M* biotin-11-dUTP (Sigma), 20 pmol T3 and T7 sequencing primers, 2.5 U Amplitaq (Perkin Elmer, Cetus) and 50 ng pooled microclone DNA. The amplification procedure was as follows: initial denaturation for 3 min at 94°C, 30 cycles of 20 s at 94°C, 20 s at 56°C, 30 s at 72°C, and final extension for 10 min at 72°C. PCR-labelled DNA was ethanol precipitated together with a 50-fold excess of human placenta DNA (Sigma), salmon sperm DNA (Sigma) was added to get a final concentration of 1 mg/ml. The DNA precipitate was dried and dissolved in hybridization mix (50% formamide, 2 × SSC, 50 m*M* Na phosphate pH 7.0, 10% dextran sulphate). DNA was denatured for 5 min at 75°C, cooled on ice for 1 min and allowed to reanneal 15–20 min at 37°C. Fluorescence in situ hybridization (FISH) was performed according to the protocol of Kievits et al. (1990) with some minor modifications. Briefly, chromosome preparations were treated with RNase A (0.1 mg/ml in 2 × SSC, 1 h at 37°C), 0.01% pepsin (in 0.01 *M* HCl, 10 min at 37°C) and fixed in 1% formaldehyde/50 m*M* $MgCl_2$/PBS for 10 min before dehydration through an ethanol series. Subsequently, the chromosomal DNA was denatured in 70% formamide/2 × SSC/50 m*M* sodium phosphate pH 7 under a coverslip for 2–3 min on a hot plate (80°C). Renaturation was prevented by subsequent washing in cold 70% ethanol, followed by dehydration.

The probe mixture was applied to the denatured chromosomes on prewarmed slides (37°C) and covered with coverslips. The coverslips were sealed with rubber cement. Hybridization took place overnight at 37°C in a humid chamber. Following hybridization, free probe was removed by washing 3 × 5 min in 50% formamide/2 × SSC pH 7 (45°C) and 3 × 5 min in 2 × SSC (45°C). Protein binding sites were blocked by incubation with 5% nonfat dry milk in 4 × SSC (NFDM buffer). Bound probe was visualized using avidin^{FITC} (vector) and goat-α-avidin^{Biotin} (vector) diluted in NFDM buffer. All incubations were at room temperature for 20 min. Every incubation was followed by 3 × 5 min washes in 4 × SSC/0.05% Tween-20. In total, three layers of avidin^{FITC} were applied. Finally, chromosome preparations were briefly rinsed in 4 × SSC and thereafter in 1 × PBS for 2 × 5 min. Subsequently, the preparations were dehydrated in an ethanol series (70%, 90%, 100%). Antifade medium containing DABCO (2 g/100 ml, Sigma) and propidium iodide (0.5 μg/ml, Sigma) as a counterstain dissolved in 9 parts glycerol and 1 part 0.2 *M* Tris-HCl pH 8 was added 30 min before microscopic evaluation. For microscopic analysis, a Zeiss Axiophot fluorescence microscope was used. Data were recorded on Scotch (3 M) 640 asa film.

3 Results and Discussion

3.1 Construction of a Microclone Library Specific for Human Chromosome Region 12q13–15

GTG-banded metaphase chromosome preparations from cultures of cytogenetically normal cells from human amniotic fluid were used for the microdissection of region q13–15 from chromosome 12. This region was dissected from 20 chromosomes 12, and the chromosome fragments were further processed as described before (Lüdecke et al. 1989, 1990; Senger et al. 1990). DNA from the chromosome fragments was amplified as described in Sect. 2. Briefly, DNA was digested with *Rsa*I, ligated to DNA of a pUC-derived plasmid vector ("*Sma*I *vector*") digested with *Sma*I, and amplified by the polymerase chain reaction (PCR) primed with the M13/pUC sequencing and reverse sequencing primers (Lüdecke et al. 1990). Amplified DNA fragments were subcloned into pT7T3-18U, and transformation of DH5α resulted in a microclone libary, designated ML12q1315, which consisted of about 20000 colonies. Using β-galactosidase hydrolysis of the colourless substrate X-gal as a marker to discriminate between nonrecombinant (blue colonies) and recombinant (white colonies) clones, it was estimated that about 80% of the clones were recombinant microclones. To obtain an impression of the average size of the DNA inserts of microclone library ML12q1315, DNA of randomly picked microclones was isolated, and upon PCR amplification DNA inserts were analysed (Fig. 1). The insert size of 30 microclones varied between 130 and 360 bp with an average of about 210 bp (corrected for the extra size caused by the T3 and T7 primers and the multiple cloning site of pT7T3-18U). This size is in good correspondence with earlier published data (MacKinnon et al. 1990; Fiedler et al. 1991).

To determine the subchromosomal location of the DNA inserts of the microclones, FISH analysis was performed. As molecular probe in such chromosome painting experiments, PCR-amplified DNA of the entire microclone library was used, as described in Sect. 2. From G-banding and chromosome painting analysis of 30 metaphase spreads of normal human lymphocytes it was concluded that only chromosome region 12q13–15 was

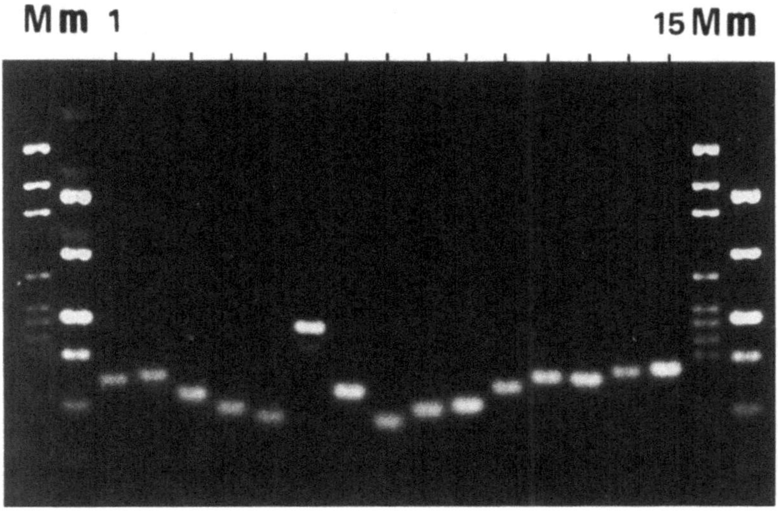

Fig. 1. Size analysis of the DNA inserts of randomly picked clones of microclone library ML12q1315. Molecular weight markers shown in *lane M* are DNA fragments generated by digestion of pGEM-3 DNA with a combination of the restriction endonucleases HinfI, *Rsa*I and *Sin*I; those in *lane m* are DNA fragments of pKUN19 digested with *Taq*I

intensely stained, in all cases. Some variations in the broadness of the stained region was observed, but this seemed to correlate with the degree of condensation of the chromosomes in the metaphase spreads. As an illustration, a typical example of the staining pattern is shown in Fig. 2. As can be seen in the figure, there was no detectable staining of other chromosomes, excluding major contamination by DNA from other chromosomes during the chromosome microdissection and DNA amplification procedure. In conclusion, the results clearly indicate that the DNA inserts of this microclone library represent a large collection of markers for region q13–15 of human chromosome 12.

3.2 FISH Analysis of Solid Tumours with Aberrations Involving Chromosome 12q13–15

With microclone library ML12q1315 chromosome painting experiments were performed to determine the relative positions of the chromosome 12 breakpoints in four solid tumours with translocations involving region q13–15 of chromosome 12. Solid tumours under investigation here included myxoid liposarcoma, lipoma, uterine leiomyoma and pleomorphic adenoma of the salivary gland. Instead of primary tumour cells, cell lines developed from primary tumours were used in these studies because their mitotic indices were much higher than those usually observed in cultures of the

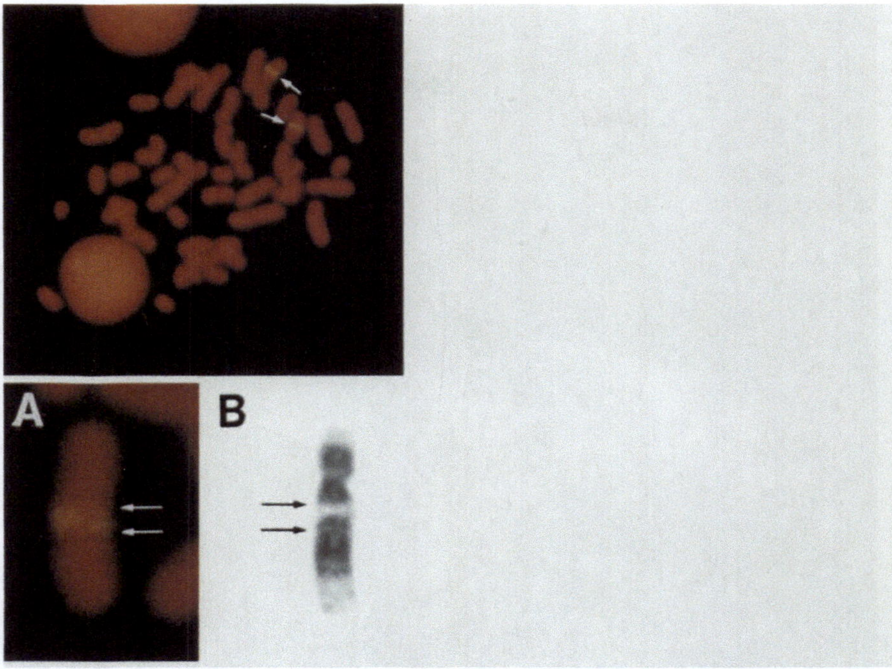

Fig. 2. Staining pattern of chromosome 12 in a metaphase spread of a human lymphocyts using the DNA inserts of microclone library ML12q1315 for chromosome painting. **A** Enlargement of a painted chromosome 12. **B** G-banded chromosome 12

corresponding primary tumours. The cell lines used included myxoid liposarcoma cell line LIS-3/SV40 (Schoenmakers et al., this volume), lipoma cell line LI-14/SV40 (Belge et al. 1992), uterine leiomyoma cell line LM-30.1/SV40 (Stern et al. 1991), and cell line PA-312 which was derived from a pleomorphic adenoma of the salivary gland (Kazmierczak et al. 1990). These cell lines were obtained after the introduction of genetic sequences of the "early region" of the SV40 virus genome into the primary tumour cells (Kazmierczak et al. 1990; Stern et al. 1991). Lipoma cell line LI-14/SV40 is characterized by a t(3;12)(q28;q14), the uterine leiomyoma cell line LM-30.1/SV40 by a t(12;14)(q14–15;q24), cell line PA-312 by a t(1;12)(p21;q14), and cell line LIS-3/SV40 by a t(12;16)(q13;p11.2). Figure 3 presents an example of a chromosomes painting experiment of a metaphase spread of LM-30.1/SV40 cells using ML12q1315 microclone DNA as probe; the corresponding G-banding pattern is also shown. The results reveal painting of region q13–15 of chromosome 12 and small regions of the derivative chromosomes der(12) and der(14), indicating that the microclone DNA inserts map on both sides of the breakpoint in chromosome 12. Results of similar chromosome painting studies of metaphase spreads also of other cell lines are given in Fig. 4. As can be seen in Fig. 4A, three chromosomes in the

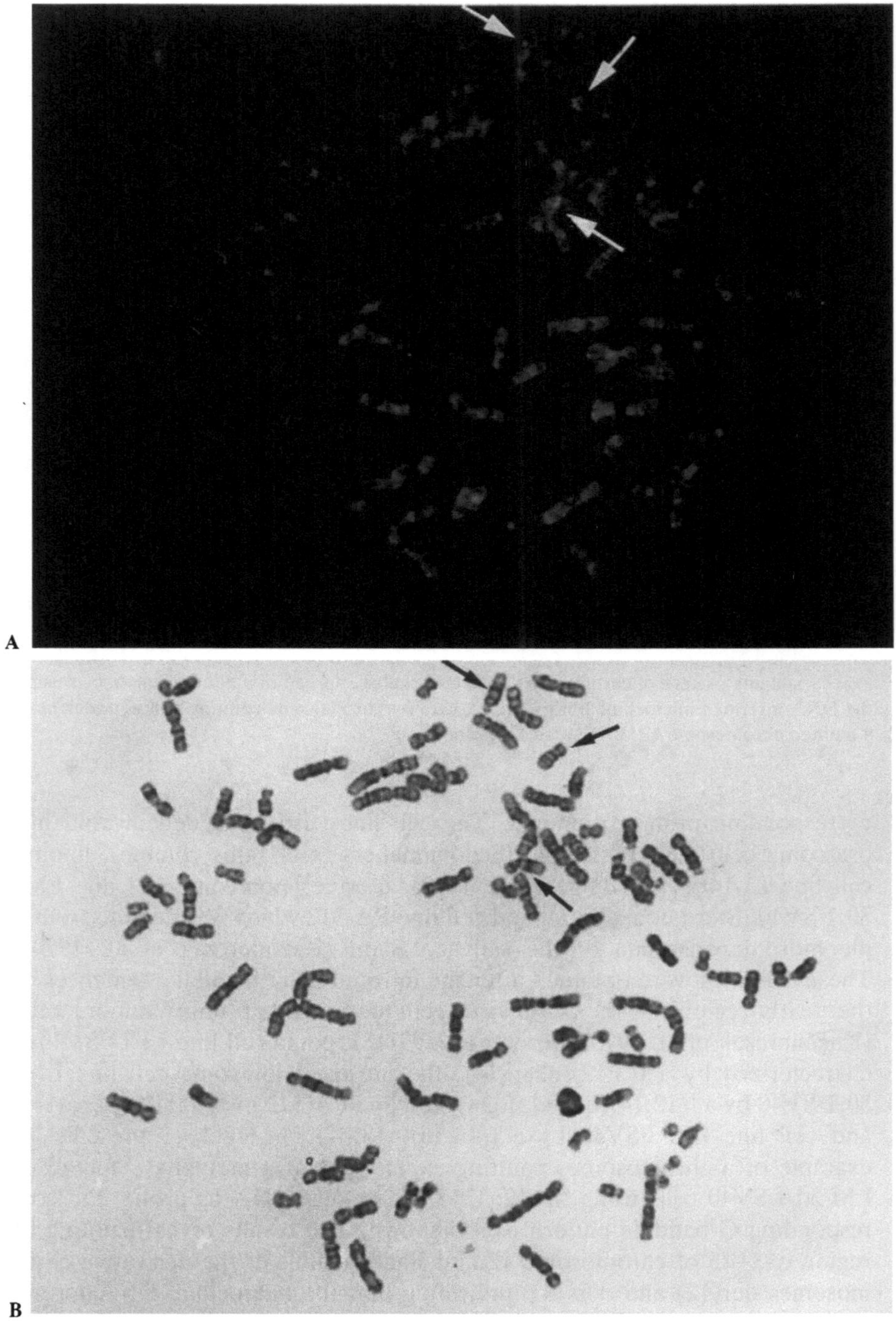

Fig. 3. A Simultaneous G banding and painting analysis of the t(12;14)(q14–15;q24) in cell line LM-30.1/SV40. Normal chromosomes 12, the derivative chromosomes der(12) and der(14) are indicated by arrows. **B** G-banded metaphase spread of the same metaphase

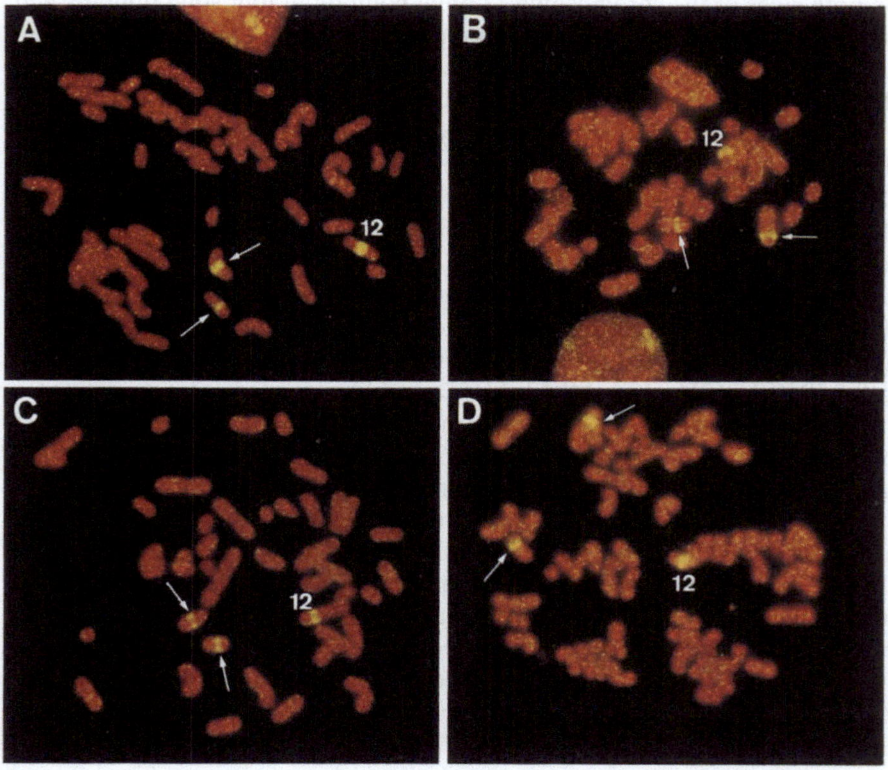

Fig. 4. Chromosome painting analysis of metaphase chromosomes of the myxoid liposarcoma cell line LIS-3/SV40 (**A**), the lipoma cell line LI-14/SV40 (**B**), the uterine leiomyoma cell line LM-30.1/SV40 (**C**), and cell line PA-312 which was derived from a pleomorphic adenoma fo the salivary gland (**D**). Normal chromosomes 12 are indicated by its number. *Arrows*, the various derivative chromosomes

metaphase spread of the myxoid liposarcoma cell line LIS-3/SV40 are stained. G-banding indicated that this staining was on normal chromosome 12 (region q13–15), der(12) and der(16). Comparative analysis of the staining patterns of der(12) and der(16) in 30 metaphases revealed that most of the staining was always found on der(16), although minor fluctuations were observed; such fluctuations might be due to differential chromosome condensation. Similar analysis of metaphase spreads of cells of lipoma cell line LI-14/SV40 (Fig. 4B), uterine leiomyoma cell line LM-30.1/SV40 (Fig. 4C) and the PA-312 cell line of a pleomorphic adenoma of the salivary gland (Fig. 4D) also revealed staining of both derivative chromosomes. Unlike in the myxoid liposarcoma cell line, however, the stained regions on the two derivative chromosomes in the cell lines of the three benign tumours appeared to be more or less similar in size. The results of these molecular studies indicate that the DNA inserts of microclone library ML12q1315 map on both sides of the chromosome 12 breakpoint of all four solid tumours and suggest that the chromosome 12 breakpoint in myxoid liposarcoma cell line LIS-3/SV40 is proximal to those in the cell lines of the three benign tumours; the latter is in support of cytogenetic data obtained before.

To characterize the somatic cell hybrid LIS-3/SV40/A9-B4, microclone library ML12q1315 was also used in a painting experiment (data not shown). the LIS-3/SV40/A9-B4 hybrid cell line was obtained upon fusion of cells of the myxoid liposarcomas cell line LIS-3/SV40 to mouse A9 cells. Of importance in the context here is to note that in this cell hybrid, derivative chromosome der(16) is retained and segregated from derivative chromosome der(12) and the normal human chromosomes 12 and 16 (Schoenmakers et al., this volume). Chromosome painting analysis of metaphase spreads of LIS-3/SV40/A9-B4 using ML12q1315 as probe revealed staining only of a small region of der(16). G banding indicated that this region was derived from the long arm of chromosome 12 and corresponded to the area immediately distal to the 12q13 breakpoint.

Finally, metaphase spreads of a well-differentiated liposarcoma and an atypical lipoma were also studied in chromosome painting experiments with the ML12q1315 microclone library as probe. The liposarcoma was resected from a 50-year-old male patient. Cytogenetic analysis of metaphase spreads revealed that the cells could be divided on the basis of their karyotypes in two categories; one had the karyotype 47,XY,+R and the other was tetraploid and had ring and long marker chromosomes. It is well established that well-differentiated liposarcomas are often characterized by giant marker chromosomes, ring chromosomes, and telomeric associations. The atypical lipoma was resected from a male patient 67 years old. The karyotype of the tumour cells was 46,XY,+R. FISH analysis of metaphase spreads of tumour cells of both patients revealed extensive but discontinuous staining throughout the entire ring chromosomes (Fig. 5). The results indicated that genetic sequences from region q13–15 of chromosome 12 are amplified in these tumours. It is known that region q13–15 of chromosome 12 contains genetic sequences, e.g. GLI1 and SAS, which are amplified in particular solid tumours (Kinzler et al. 1987; Meltzer et al. 1991). At present, the nature of the amplified genetic sequences in the ring chromosomes remains to be established.

The results of this study clearly establish that microclone library ML12q1315 has specificity for region q13–15 of human chromosome 12 and that the DNA inserts of it map on both sides of the chromosome 12 breakpoints observed in myxoid liposarcoma, lipoma, uterine leiomyoma and pleomorphic adenoma of the salivary gland. The availability of such a collection of microclones is of importance for future mapping studies of region q13–15 of human chromosome 12, and the library is expected to facilitate molecular characterization of chromosome breakpoints in this region. The microclone library is also expected to be instrumental in conclusively answering the question as to whether the same or different genetic

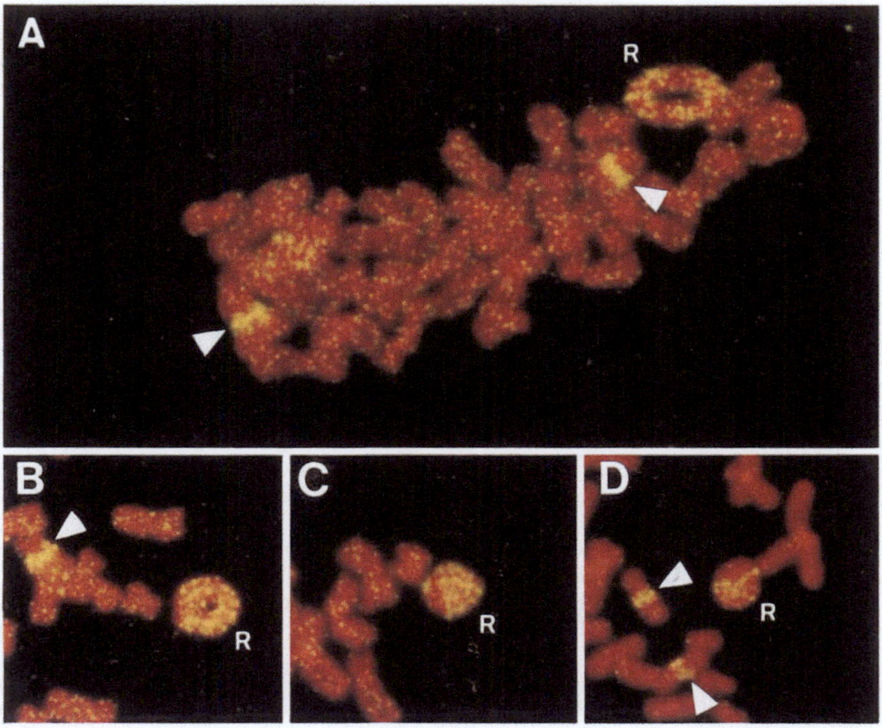

Fig. 5. A Chromosome painting patterns of ring chromosomes in a metaphase spread from a well-differentiated liposarcoma using microclone library ML12q1315 as probe. *R*, A painted ring chromosome; *arrowheads*, painted chromosomes 12. **B, C, D** Other examples of painted chromosomes 12 (*arrowheads*) and ring chromosomes (*R*) from the well-differentiated liposarcoma

sequences are affected by the chromosome 12 breakpoints in myxoid liposarcoma and the three benign tumours. The first molecular mapping data obtained with the newly constructed ML12q1315 microclone library point towards differences in the breakpoints.

Acknowledgements. The authors with to thank L. Mekers, R. Mols and K. Meyer-Bolte for their excellent technical assistance and S. Bartnizke, P. Dal Cin and M. Dehaen for cytogenetic analysis. This work was supported in part by the EC Medical and Health Research Programme, the EC-sponsored concerted action program "Molecular Cytogenetics of Solid Tumours", by the Inter-University Network for Fundamental Research sponsored by the Belgian Government (1991–1995), the "Geconcerteerde Onderzoekacties 1992–1996" and the Schwerpunktprogramm "Molekulare und Klassische Tumorcytogenetik" of the Deutsche Forschungsgemeinschaft.

References

Belge G, Kazmierczak B, Meyer-Bolte K, Bartnitzke S, Bullerdiek J (1992) Expression of SV40 T antigen in lipoma cells with the chromosomal translocation t(3;12) is not sufficient for direct immortalization. Cell Biol (Int Rep) 16:4

Fiedler W, Claussen U, Lüdecke H-J, Senger G, Horsthemke B, Geurts Van Kessel A, Goertzen W, Fahsold R (1991) New markers for the neurofibromatosis-2 region generated by microdissection of chromosome 22. Genomics 10:786–791

Kazmierczak B, Bartnitzke S, Hartl M, Bullerdiek J (1990) In vitro transformation by the SV40 "early region" of cells from a human benign salivary gland tumour with a 12q13 → q15 rearrangement. Cytogenet Cell Genet 53:37–39

Kievits T, Dauwerse JG, Wiegant J, Devilee P, Breuning MH, Cornelisse CJ, van Ommen G, Pearson PL (1990) Rapid subchromosomal localization of cosmids by nonradioactive in situ hybridization. Cytogenet Cell Genet 53:134–136

Kinzler KW, Bigner SH, Bigner DD, Trent JM, Law ML, O'Brien SJ, Wong AJ and Vogelstein B (1987) Identification of an amplified, highly expressed gene in a human glioma. Science 236:70–73

Lüdecke H-J, Senger G, Claussen U, Horsthemke B (1989) Cloning defined regions of the human genome by microdissection of banded chromosomes and enzymatic amplification. Nature 338:348–350

Lüdecke H-J, Senger G, Claussen U, Horsthemke B (1990) Construction and characterization of band-specific DNA libraries. Hum Genet 84:512–516

MacKinnon RN, Hirst MC, Bell MV, Watson JEV, Claussen U, Lüdecke H-J, Senger G, Horsthemke B, Davies KE (1990) Microdissection of the fragile X region. Am J Hum Genet 47:181–187

Melzer PM, Jankowski JA, Dal Cin P, Sandberg AA, Paz IB, Coccia MA (1991) Identification and cloning of a novel amplified DNA sequence in human malignant fibrous histiosarcoma derived from a region of chromosome 12 frequently rearranged in soft tissue tumors. Cell Growth Differ 2:495–501

Mitelman F (1991) Catalog of chromosome aberrations in cancer, 3rd edn. Liss, New York

Ohjimi Y, Iwasaki H, Kaneko Y, Ishiguro M, Ohgami A, Fujita C, Shinohara N, Yoshitake K, Kikuchi M (1992) Chromosome abnormalities in liposarcomas. Cancer Genet Cytogenet 64:111–117

Pandis N, Heim S, Willen H, Bardi G, Floderus UM, Mandahl N, Mitelman F (1991) Chromosome analysis of 96 uterine leiomyomas. Cancer Genet Cytogenet 55:11–18

Sandberg AA (1990) The chromosomes in human cancer and leukemia. Elsevier, New York

Sandros J, Stenman G, Mark J (1990) Cytogenetic and molecular observations in human and experimental salivary gland tumours. Cancer Genet Cytogenet 44:153–167

Senger G, Lüdecke H-J, Horsthemke B, Claussen U (1990) Microdissection of banded human chromosomes. Hum Genet 84:507–511

Sreekantainah C, Leong SLP, Karakousis CP, McGee DL, Rappaport WD, Villar HV, Neal D, Fleming S, Wankel A, Herringto PN, Carmona R, Sandberg AA (1991) Cytogenetic profile of 109 lipomas. Cancer Res 51:422–433

Stern C, Kazmierczak B, Thode B, Rommel B, Bartnitzke S, Dal Cin P, Van de Ven W, Van den Berghe H, Bullerdiek J (1991) Leiomyoma cells with 12q15 aberrations can be transformed in bvitro and show a relatively stable karyotype dring precrisis period. Cancer Genet Cytogenet 54:223–228

Turc-Carel C, Limon J, Dal Cin P, Rao U, Karakousis C, Sandberg AA (1986a) Cytogenetic studies of adipose tissue tumours. II. Recurrent reciprocal translocation t(12;16)(q13;p11) in myxoid liposarcomas. Cancer Genet Cytogenet 23:291–299

Turc-Carel C, Dal Cin P, Rao U, Karakousis C, Sandberg AA (1986b) Cytogenetic studies of adipose tissue tumours. I. A benign lipoma with reciprocal translocation t(3;12)(q28;q14). Cancer Genet Cytogenet 23:283–289

Isolation and Characterisation of *Not*I-end Cosmids Mapping to Human Chromosome 12p

J. Zhang, M. Baens, M. Chaffanet, J. Aerssens, J.-J. Cassiman, and P. Marynen

1 Introduction

The short arm of human chromosome 12 represents approximately 1.5% of the human genome. The cytogenetic analysis of 12p by classical banding techniques is therefore limited to the detection of quantitative or major structural alterations.

Human germ cell tumors occur with a wide clinical and histopathological variety (see de Jong et al. 1990). Despite this varying histological appearance, germ cell tumors show distinct cytogenetic aberrations, the most prominent structural abnormality being the presence of an isochromosome 12p in the majority of the gonadal germ cell tumors (Delozier-Blanchet et al. 1985; Castedo et al. 1988). Cytogenetic data on extragonadal germ cell tumors are scarce due to the low incidence of those tumors, but the available data point to the presence of an i(12)p as the sole chromosomal abnormality in a majority of these tumors (Dal Cin et al. 1989). Also, in some tumors without i(12)p other structural changes of 12p were detected with fluorescent in situ hybridisation (Suijkerbuijk et al. 1991). The hypothesis can therefore be formulated that gene(s) located on chromosome 12p are specifically involved in the genesis or the neoplastic evolution of germ cell tumors.

It is our general aim to develop tools for the molecular analysis of chromosome 12p. Different strategies are developed to isolate unknown 12p genes. The isolation and partial characterisation of *Not*I-end cosmids mapping to 12p is reported here. The cosmids must allow the structural characterisation of 12p by fluorescent in situ hybridisation. In addition, since the clones are derived from *Not*I sites and thus likely to be derived from CpG islands, which are known to be associated with expressed sequences, the cosmids will be useful for the identification of new 12p genes.

Center for Human Genetics, University of Leuven, 3000 Leuven, Belgium

2 Materials and Methods

2.1 General Methods

The establishment and maintenance of the M28 somatic cell hybrid and of the different cell lines was as reported before (Zhang et al. 1989). All DNA labelling, Southern and northern analyses were performed according to standard procedures, unless otherwise indicated. Sequencing was performed using fluorescent primers on an ALF sequencer (Pharmacia, Uppsala, Sweden). Analysis of the sequences was performed using the Intelligenetics Suite (Intelligenetics, Mountain View, CA).

2.2 Isolation of *Not*I-end Clones

The cosmid vector pWE15 (Stratagene, La Jolla, CA) was digested with *Not*I and *Cla*I and purified by agarose gel chromatography. An adaptor with the appropriate cohesive ends and an internal *Bam*HI site was synthesized, kinased and ligated into the digested vector. The presence of unique *Not*I and *Bam*HI sites in the final construct was confirmed by restriction analysis.

High molecular weight DNA from M28 cells was prepared in low melting point agarose blocks as described by van Ommen and Verkerk (1986). The DNA in the agarose blocks was completely digested with *Not*I. The blocks were then equilibrated with 2 × *Sau*3A buffer, melted at 65°C and cooled to 37°C. The DNA was then partially digested with *Sau*3A. The DNA was size fractionated on a sucrose gradient (10%–40% sucrose in 1 *M* NaCl, 20 m*M* Tris-HCl pH 8.0, 5 m*M* EDTA) in an SW28 rotor at 26000 rpm for 24 h at 20°C. Fractions containing DNA ranging from 30 to 40 kb were dialysed, and the DNA was ethanol precipitated.

Insert DNA was ligated into *Not*I-BamHI digested vector, packaged and transformed into the *Escherichia coli* strain CPLK (Stratagene, La Jolla, CA). To increase the yield of cosmid DNA preparations, the cosmids were subsequently transferred to the *E. coli* NM522 strain.

2.3 Fluorescence In Situ Hybridisation

Prometaphase chromosome spreads were obtained according to Yunis (1981) from lymphocyte cultures as follows. The cells were synchronised by adding $10^{-7}\,M$ methotrexate to the medium and released by adding BrDU ($6.5 \times 10^{-5}\,M$). The cell cycle was arrested at the prometaphase by a short exposure of the cultures to colcemid (0.1 μg/ml). The cells were swollen with 0.075 *M* KCl, fixed with methanol/glacial acetic acid (3/1), dropped on slides aged for 2–10 days at room temperature. Fluorescent in situ hybridisation with biotinylated cosmid DNA was performed as described by Kievits et al. (1990).

3 Results

3.1 Isolation of *Not*I-end Clones

The isolation of high molecular weight DNA is essential for the generation of a representative cosmid library. Therefore the protocol originally developed for the generation of DNA for pulsed field gel electrophoresis was adapted. M28 cells were imbedded in agarose and treated with proteinase K as described (van Ommen and Verkerk 1986). The DNA was then digested extensively in the agarose with *Not*I. Both the integrity of the DNA before digestion and the *Not*I digestion were checked by pulsed field gel electrophoresis (not shown). The DNA-agarose blocks were then melted at 65°C, and the optimal enzyme concentration for partial digestion with *Sau*3A was determined. After digestion, the DNA fragments were fractionated on a sucrose gradient (see Sect. 2), and the size fraction between 30 and 50 kb was ligated into a cosmid vector modified to bear cohesive *Not*I and *Bam*HI ends. About 10 000 colonies, obtained from 1 μg insert genomic DNA, were plated in duplicate onto nitrocellulose filters and hybridised to total human genomic DNA. Ninety-three colonies were positive, which is proportional to the estimated human DNA content of the M28 cell line. After a second round of screening 48 clones hybridising exclusively with human DNA were recovered, while 30 clones were shown to hybridise to both human and mouse total genomic DNA. Eighteen clones were then purified for further analysis.

First a restriction analysis was performed with *Not*I (GCGGCCGC), *Eag*I (CGGCCG), *Sac*II (CCGCGG) and *Bss*HII (GCGCGC). As shown in Table 1 all clones contain sites for these enzymes, further illustrating their origin from CpG rich stretches of the human genome. While 16 clones showed one *Not*I site, two showed two *Not*I sites. This could either be the result of an incomplete *Not*I digestion of the M28 DNA or reflect methylation of the second *Not*I site in the M28 cell line.

3.2 Detection of Two 12p Genes

As a first approach to detect genes eventually associated with the CpG rich regions cloned into the cosmids, total cosmid DNA was labelled, denatured and preannealed with total human DNA. The labelled probe was then hybridised to northern blots carrying different RNA samples. Three clones showed positive signals (Fig. 1A,B) Two clones, A8 and H, detected a

Table 1. Restriction analysis of the cosmid clones

Restriction sites for clone	*Not*I GCGGCCGC	*Eag*I CGGCCG	*Sac*II CCGCGG	*Bss*HII GCGCGC
A	1	1	2	2
A2	1	0	1	3
A3	1	1	0	1
A8	1	2	8	3
A9	1	1	1	1
A22	1	4	4	3
A26	2	2	3	4
A29	1	0	1	0
7A	1	3	3	3
13A	1	0	2	1
17A	1	1	2	1
D	1	4	4	1
E	1	1	1	2
H	1	1	5	2
J	1	0	1	0
L	1	0	1	1
P	2	1	3	2
Y	1	0	2	1

DNA of the cosmids was digested with the enzymes indicated and separated on 0.3% agarose gels. The recognition site for *Eag*I is contained within the site for *Not*I; however, only additional *Eag*I sites were counted.

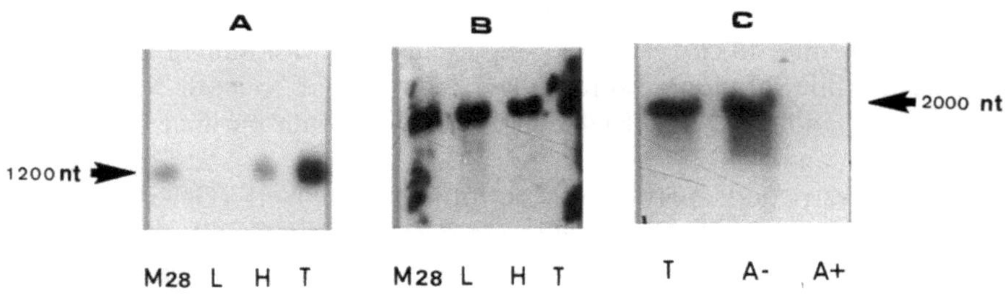

Fig. 1A–C. Northern analysis. RNA isolated from human HepG2 cells (*H*), M28 cells (*M28*), Tera2 cells (*T*), and LTk$^-$ (*L*) cells was separated on denaturing agarose gels and transferred to nylon membranes. **A** Hybridisation with total cosmid DNA from clone A8 (GAPDH). **B** Hybridisation with clone A9. **C** Total RNA (*T*), poly(A)$^-$ (*A*−) and poly(A)$^+$ (*A*+) RNA was isolated from a teratoma cell line, blotted and hybridised with a 636-bp fragment of clone A9

mRNA or 1.2 kb on northern blots with RNA from HepG2 cells, M28 cells and Tera2 cells. No signal was observed with RNA derived from murine LTKcells (Fig. 1A). For both A8 and H a 900-bp single-copy subclone was sequenced. The sequence was compared to the sequences in the EMBL databank and turned out to be identical to part of the glyceraldehyde-3-phosphate dehydrogenase gene (Ercolani et al. 1988) previously mapped on chromosome 12p13 (see Craig and McBride 1991). Although both A8 and H start from the same *Not*I site, the inserts are of different length (A8, 40 kb; H, 28 kb) which explains the different number of restriction sites observed with rare cutter enzymes, as indicated in Table 1.

A third cosmid, A9, detected a RNA of approximately 2 kb, present in total RNA from both human and murine origin (Fig. 1B). A 636 bp subclone of A9 was isolated showing the same hybridisation pattern. Since no signal was observed with poly(a)$^+$ RNA, the poly(A)$^-$ fraction not binding to an oligo(dT) column was collected and used in a northern experiment. Figure 1C shows that an equally strong signal was obtained with the RNA fraction not binding to an oligo(dT) column compared to total RNA. Taken together this indicates that clone A9 detects a highly conserved, nonpolyadenylated RNA. Sequence analysis however detected no significant homology with any sequence present in the EMBL database (release 30) which seems to exclude the presence of a gene for ribosomal RNA on cosmid A9.

3.3 Regional Assignment of the Cosmid Clones

All 18 cosmid clones were mapped by fluorescence in situ hybridisation on prometaphase spreads obtained from lymphocyte cultures. Figure 2 shows the regional assignment of the 18 cosmids as scored by two independent investigators. The regional mapping was facilitated by the fact that under the experimental conditions used, many mitoses revealed *Alu* banding

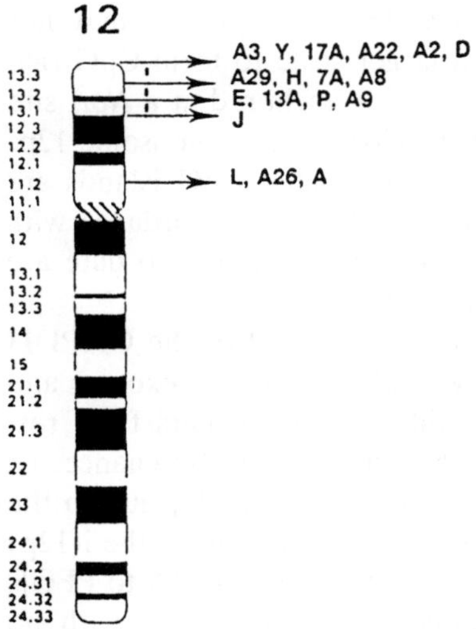

Fig. 2. Regional assignment of the clones by fluorescent in situ hybridisation. Biotinylated cosmid DNA was hybridized to human prometaphase chromosomes. The slides were developed with FITC avidin and biotinylated anti-avidin antibodies

patterns resembling an R-banding pattern (Korenberg and Rykowski 1988; Lichter et al. 1990). It is interesting to note that the cosmids clearly cluster on chromosome 12p, 15 of them mapping to 12p13 while the remaining three cosmids mapped to 12p11, the second light staining region of 12p upon G banding.

4 Discussion

Using DNA from the hybrid cell line M28 a cosmid library with clones anchored at one side to a *Not*I restriction site was derived. Eighty cosmids hybridising to total human genomic DNA were isolated. A number of clones showed cross-hybridisation with total human and mouse genomic DNA. Although this could also be due to the presence of evolutionary conserved sequences (Burton et al. 1986), 18 clones hybridising exclusively to human DNA were chosen for further experiments.

All clones were shown to be derived from CpG rich regions of 12p. The 12p origin of all the cosmids further demonstrates that the isochromosome 12p is the only human DNA present in the M28 cell line further extending the observations of Zhang et al. (1989) and of Suijkerbuijk et al. (1991). Although no bias other than the presence of a *Not*I site was intended, mapping of the cosmids by fluorescence in situ hybridisation clearly indicated the majority of the cosmid clones to be localised on 12p13, with three clones mapping on 12p11. Holmquist et al. (1982) showed that Giemsa dark bands correspond to AT-rich regions while GC-rich regions would result in Giemsa light bands. It has also been suggested that CpG rich sequences may not only form CpG islands but may cluster on a larger scale to form CpG-rich isochores (Bernardi 1989). The isolation of cosmids anchored at a *Not*I site might thus explain the regional clustering of the clones on chromosome 12p. This could, however, also indicate that undermethylated CpG islands are more abundant in these regions of 12p, which would be in accordance with the observation that the majority of the genes mapped on 12p to date are found in the p13 region (Craig and McBride 1991).

Sequence analysis showed two cosmids to contain part of the GAPDH gene, downstream of the *Not*I site previously localised between exons 1 and 2 of the gene. Since the other cosmids are different as appears from restriction analysis, it is not clear wether this occurred purely by chance, or whether the complexity of the library is very low, for example, due to the presence of only a limited number of unmethylated *Not*I sites on the i(12p) chromosome of M28. Hybridisation of total human genomic DNA to PFGE blots of *Not*I digested DNA, however, stained the whole lane, with no distinct bands visible (not shown), arguing against this hypothesis. The detection of the GAPDH gene by blotting of total cosmid DNA to northern blots indicates this to be a valid approach to identify eventual genes in the different cosmids. The absence of any specific signal as observed with the other cosmids could indicate either the absence of sufficient coding sequences in those cosmid or the absence of the corresponding mRNA in the source of RNA used. Hybridisation selection of cDNA and subsequent PCR might be a valid alternative in this regard (Parimoo et al. 1991).

The detection of a poly(A)$^-$ RNA by a 636-bp fragment of A9 is not yet explained. This is thought not to be an artefact resulting from the partial degradation of the RNA preparation since northern analysis of the poly(A)$^+$ RNA from the same preparation with the GAPDH probe showed no signs of degradation, while neither the GAPDH probe nor an other cDNA coding for triosephophate isomerase 1 revealed any signal on the northern blots with the poly(A)$^-$ RNA (not shown). Experiments by Van Ness et al. (1979) suggested that RNA isolated from mouse brains contained a non-polyadenylated RNA population with a complexity similar to the population of polyadenylated RNA population on the one hand, and that the two populations shared very few sequences with each other. Only a few genes coded for by poly(A)$^-$ RNA are known however (Albig et al. 1991), and a similarity search of the EMBL databank revealed no significant homologies. We do not know yet, however, which fragment of the subclone hybridises to the RNA, and the isolation of the corresponding cDNA will allow us to analyse the eventual coding potential of this RNA.

In conclusion, 17 different *Not*I-end were isolated and mapped to chromosome 12p. Two cosmids were shown to contain part of the GAPDH gene, one cosmid contains sequences hybridising to nonpolyadenylated RNA. We expect the clones to be useful tools for the characterisation of structural chromosome 12p anomalies by FISH. The cosmids are further analyzed for the presence of yet unknown genes and thereby complement other strategies developed by us to isolate new 12p genes.

References

Albig W, Kardanilou E, Drabent B, Zimmer A, Doenecke D (1991) Isolation and characterisation of two human H1 histone genes within clusters of core histone genes. Genomics 10:940–948

Burton FH, Loeb DD, Voliva CF, Martin SL, Edgell MH, Hutchinson CA (1986) Conservation throughout mammalia and extensive protein-encoding capacity of the highly repeated DNA long interspersed sequence one. J Mol Biol 187:291–304

Bernardi G (1989) The isochore organisation of the human genome. Annu Rev Genet 23:637–661

Castedo SMMJ, deJong B, Oosterhuis JW, Seruca R, Indenbury WJ Buist J, Sleijfer DT (1988) i(12p) negative testicular germ cell tumors a different group? Cancer Genet Cytogenet 35:171–178

Craig IW, McBride OW (1991) Report of the committee on the genetic constitution of chroomosome 12. Cytogenet Cell Genet 58:555–579

Dal Cin P, Drochmans A, Moerman P, Van den Berghe H (1989) Isochromosome 12p in a mediastinal germ cell tumor. Cancer Genet Cytogenet 42:243–251 de Jong B, Oosterhuis JW, Castedo SMMJ, Vos A, te Meerman GJ (1990) Pathogenesis of adult testicular germ cell tumors. Cancer Genet Cytogen 48:143–167

Delozier-Blanchet CD, Engel E, Wolt H (1985) Isochromosome 12p in malignant testicular tumors. Cancer Genet Cytogenet 15:375–376

Ercolani L, Florence B, Denaro M, Alexander M (1988) Isolation and complete sequence of a functional human glyceraldehyde-3-phosphate dehydrogenase gene. J Biol Chem 263: 15335–15341

Holmquist GP, Gray M, Porter T, Jordan J (1982) Characterisation of Giemsa dark- and light-band DNA. Cell 31:121–129

Kievits T, Dauwerse JG, Wiegant J, Deville P, Breuning MH, Cornelisse CJ, van Ommen G-J, Pearson PL (1990) Rapid subchromosomal localisation of cosmids by nonradioactive in situ hybridisation. Cytogenet Cell Genet 53:134–136

Korenberg JR, Rykowski MC (1988) Human chromosome organisation: alu, lines, and the molecular structure of metaphase chromosome bands. Cell 53:391–400

Lichter P, Tang CC, Call K, Hermanson G, Evans GA, Housman D, Waed DC (1990) High resolution mapping of uman chromosome 11 by in situ hybridisation with cosmid clones. Science 247:64–69

Parimoo S, Patanjali SR, Shukla H, Chaplin DD, Weissman SM (1991) cDNA selection: efficient approach for the selection of cDNA's encoded in large chromosomal DNA fragments. Proc Natl Acad Sci USA 88:9623–9627

Suijkerbuijk RF, Van de Veen AY, Van Echten J, Buys CHCM, De Jong B, Oosterhuys JW, Warburton DA, Cassiman J-J, Schonk D, Vân Kessel AG (1991) Demonstration of the genuine iso-12p character of the standard marker chromosome of testicular germ cell tumor and identification of further chromosome 12 aberrations by competitive in situ hybridisation. Am J Hum Genet 48:269–273

Van Ness J, Maxwell IH, Hahn WE (1979) Complex population of nonpolyadenylated messenger RNA in mouse brain. Cell 18:1342–1349

van Ommen GJB, Verkerk JMH (1986) Restriction analysis of chromosomal DNA in a size range up to two million base pairs by pulsed field gradient electrophoresis. In: Davies KE (ed) Human genetic diseases. A practical approach. •• pp 113–133

Yunis JJ (1981) Mid-prophase human chromosomes. The attainment of 2000 bands. Hum Genet 56:293–298

Zhang J, Marynen P, Devriendt K, Frijns J-P, Van den Berghe H, Cassiman J-J (1989) Molecular analysis of the isochromosome 12p in the Pallister-Killian syndrome. Hum Genet 83:359–363

Appendix

Selected Further Reading

Since aberrations of chromosome 12 are also frequently observed in non-solid tumors, some of these papers have been included in this list as well.

Cytogenetics and Molecular Genetics of Pleomorphic Adenomas

- Pleomorphic adenomas with 8q12- or 12q13–15-rearrangements vary in their ability to be transformed by SV40 large T-antigen (Kazmierczak et al. 1992).
- In a series of 220 salivary gland pleomorphic adenomas reported by Bullerdiek et al. (1993) 13% showed 12q13–15 involvement.
- The chromosome 12 breakpoint of one adenoma (Ad 312) as assigned to a region of less than 165 kbp. FISH evaluation of the chromosome 12 breakpoints in five other adenoma cell lines indicated that they are located proximal to the one of Ad 312 (Kools et al. 1994).

Cytogenetics and Molecular Genetics of Uterine Leiomyomas

- In a study of 93 uterine leiomyomas karyotyped by Meloni et al. (1992) correlations between the cytogenetic type and the histology of the tumors were investigated. In these benign tumors, no such correlation became apparent.
- Stern et al. (1992) reported the karyotypes of 139 uterine leiomyomas.
- A t(12;14) identical to that found as a specific abnormality in uterine leiomyomas has been described by Hu et al. (1992) in a uterine lipo-leiomyoma.
- The breakpoint of one uterine leiomyoma with a translokation t(12;14) (q15;q24) was molecularly cloned and characterized by Schoenmakers et al. (1994).

Cytogenetics of Adipose Tissue Tumors

- Belge et al. (1992) have described the successful transformation/immortalization of a lipoma showing a t(3;12)(q28;q14) by SV40 large T-antigen.
- Mitchell et al. (1992) reported the presence of ring chromosomes 11 cells were analyzed: one large ring: 5, one large and one small: 3, one large and two small: 1, normal karyotype: 2, in a retroperitoneal lipoma showing no atypical features.
- Pedeutour et al. (1993) have studied five cases of well-differentiated liposarcoma, each showing supernumerary ring or giant marker chromosomes. All supernumerary chromosomes showed involvement of chromosome 12.
- A fibrolipoma showed a t(12;16)(q13;q24) as the sole cytogenetic abnormality (Tayyeb et al. 1993). Since the same translocation has previously been reported in a lipoma, the authors concluded that fibrolipoma may be histologically, but not cytogenetically, distinct from lipoma.
- Four liposarcomas were karyotyped by Ohjimi et al. (1992). Of these, three myxoid liposarcomas showed the specific t(12;16)(q13;p11) either as the sole abnormality (two cases) or with an additional trisomy 8. The fourth tumor had complex numerical as well as structural aberrations.
- Aman et al. (1992) described a rearrangement of CHOP, a gene belonging to the CCAAT/enhancer binding protein family, as the molecular background of the t(12;16) in myxoid liposarcomas. Members of this gene family are expressed at high levels in fat and seem to be involved in adipocyte differentiation.
- Mrózek et al. (1993) confirmed the observation that the 12q13–15 breakpoints in lipomas and myxoid liposarcomas are different at the cytogenetic level.
- Gemmill et al. (1992) have isolated a YAC clone spanning the t(12;16) in myxoid liposarcomas and also containing Gli.
- A somatic cell hybrid retaining the der(16)t(12;16)(q3;p11.2) from the t(12;16) in myxoid liposarcomas has been constructed by Schoenmakers et al. (1993).
- Ring chromosomes of different sizes were reported in a retroperitoneal lipoma (Mitchell et al. 1992).
- The usefulness of chromosome studies in soft tissue tumors was revealed by the detection of the t(12;16)(q13;p11) in a poorly differentiated sarcoma speaking in favor of the tentative diagnosis of a poorly differentiated liposarcoma (Okjimi et al. 1992).
- A well-differentiated liposarcoma characterized by an additional giant marker was reported by Stephenson et al. (1992). By centromeric probes for a large number of different chromsomes it was not possible to identify its origin.

Germ Cell Tumors

- Rodriguez et al. (1992) analyzed chromosome 12 aneuploidy in interphase cells from human male germ cell tumors by FISH. A total of 72 fresh tumor samples were analyzed. The results of cytogenetic anlyses agreed with those obtained by FISH with respect to the occurrence of i(12p) and the chromosome 12 copy number.
- Cytogenetic findings in male germ cell tumors led Murty et al. (1992) to suggest that the 12q-losses resulting from i(12p) formation and deletions of the long arm represent loss of one or more candidate tumor suppressor genes whose products regulate the normal proliferation of the spermatogonial cells. By comparing germ line and tumor genotypes for loss of heterozygosity for polymorphic loci they identified the bands 12q13 and 12q22 as the sites of the putative suppressors.
- Von Eyben et al. (1992) tested the serum lactate dehydrogenase isoenzyme 1 activity in 27 adult patients with testicular germ cell tumors. The enzyme activity was found to correlate with the total number of 12p copies/cell.
- To test the possible genetic heterogeneity within testicular germ cell tumors and their precursor lesions, i.e., carcinoma in situ of the testis, interphase cytogenetics was applied to these tumors (Looijenga et al. 1993). Invasive nonseminomatous tumors showed a significantly higher chromosome 1 and 12 copy number compared with their adjacent in situ lesions.
- Chromosome 12-derived marker chromosomes were described in three male germ cell tumors lacking the i(12p) (Atkin et al. 1993). The authors concluded from the cytogenetic findings that the amplification of genes on 12p represents a significant step in the formation of germ cell tumors.
- Cell lines from human germ cell tumors were analyzed using different 12p-derived polymorphic markers. The results revealed that i(12p) in germ cell tumors is a genuine isochromosome with genetically identical arms (Sinke et al. 1993).
- The distribution of chromosome 12 and parts of it were studied in a testicular germ cell line and its cisplatin-resistant derivative by FISH (Reilly et al. 1993). Although it is not clear whether differences might be due to the resistance itself, metaphase FISH allows accurate detection of chromosome 12 material.
- Speleman et al. (1992) described a near-diploid karyotype 49,XX, +der(6) t(1;6)(q11;q22),+8, +12,+i(12p) in a mature ovarian teratoma.

  By FISH studies on disaggregated nuclei obtained from paraffin-embedded samples form pediatric granulosa-stromal cell tumors Schofield and Fletcher (1992) found trisomy 12 in three juvenile granulosa cell tumors and one malignant granulosa cell tumor and tetrasomy 12 in an ovarian thecoma.

Other Solid Tumors with Chromosome 12 Aberrations

- Hirabayashi et al. (1992) found 12q13 rearrangements as clonal abnormalities in two moderately differentiated (grade 2) chondrosarcomas. One had a t(6;12)(q25;q13) together with other abnormalities in most of the cells analyzed, the other had a t(12;19)(q13;q13) as the sole clonal abnormality.
- Karyotypic analysis of a malignant mixed Müllerian tumor of the uterus revealed two abnormal clones, both of them showing an i(12p) in addition to other abnormalities (Sreekantaiah et al. 1992).
- Two cases of clear cell sarcoma were cytogenetically analyzed by Travis and Bridge (1992). A t(12;22)(q13;q13) was found in one of them, confirming that this translocation is a specific aberration for this type of tumor.
- Cytogenetic analysis of a recurrent periosteal chrondroma revealed a karyotype with an involvement of band q13 on both chromosomes: 45, XY,der(1)t(7;12)(q32;q13,dic(12;12)(q13;q13),der(16)t(12;16)(q13;q22). (Mandahl et al. 1993a). Changes of 12q13–15 seem to occur nonrandomly in chondromas.
- Karyotypes of four hemangiopericytomas were reported by Mandahl et al. (1993b). Three tumors had clonally acquired chromosome aberrations: inv(12)(q14q24) in a malignant hemangiopericytoma, a supernumerary der(3)t(3;12)(p21–23;q13–15) in a benign hemangiopericytoma, and a t(6;12;19)(p21;q13;p13) in a spindle cell sarcoma.
- Dal Cin et al. (1993) described an inv(12)(p13q14–15) as the sole abnormality in a pulmonary chondroid hamartoma.
- A short-term culture from a clear-cell sarcoma of tendons and aponeuroses was cytogenetically analyzed and revealed a complex karyotype including a t(12;22)(q13;q12.2–12.3) and trisomy 8 (Mrózek et al. 1992).
- Kato et al. (1993) established a small-cell lung carcinoma cell line with a complex karyotype, including a t(12q;17p).
- Cell lines form gonadal and extragonadal human germ cell tumors (GCT) and primary GCTs were tested for the presence of i(12p) and other aberrations involving 12p by FISH. The results suggest that the occurrence of the latter aberrations is also a frequent finding in GCT (Suijkerbuijk et al. 1992).
- A clear-cell sarcoma showing a t(12;22)(q13;q13) and multiple copies of chromosome 8 in addition to other abnormalities is reported by Rodriguez et al. (1992).
- Two primary benign epithelial breast tumors showing translocations involving 12q13–15 were reported by Rohen et al. (1993).

Chromosome 12 Changes in Leukemia and Lymphoma

- A dicentric translocation chromosome involving the short arms of chromosomes 9 and 12 was reported by Mahmend et al. (1992) in 15 of 2303 cases of ALL in children and is associated with an excellent prognosis.
- FISH was applied to the detection of trisomy 12 in interphase cells of seven patients with B-CLL (Qumsiyeh and Tharapel 1992).
- The translocation t(12;17)(p13;q21), an uncommon 12p abnormality of childhood acute lymphoid leukemia was identified in five of 2620 cases (0.2%) (Krance et al. 1992).
- A karyotype 47,XX,+8,t(12;22)(p13;q13) was reported in a patient with acute myelomegakaryoblastic leukemia (Kashimura and Minamihisamatin 1993). A similar translocation had been reported previously in two patients with ANLL.
- Based on two new cases of a myeloproliferative disorder with a t(5;12)(q31;p12), Wessels et al. (1993) have concluded that this translocation may represent a subgroup of patients with features of both chronic myeloid leukemia and chronic myelomonocytic leukemia.
- Döhner et al. (1993) have applied FISH for the detection of trisomy 12 in interphase cells of 45 patients with chronic lymphoid leukemias. In three patients they found evidence for trisomy 12 not detected by conventional G-banding analysis.
- By FISH studies Raghoebier et al. (1992) were able to show that in chronic lymphocytic leukemia trisomy 12 occurs in most instances only in a subpopulation of leukemia cells, suggesting that trisomy 12 is a late event in the course of CLL.
- Based on conventional G-banded preparation (475 patients) and interphase FISH (117 patients) studies in CLL Escudies et al. (1993) have compared the incidence of trisomy 12 detected by both methods. They found that only 34.6% of the cases of trisomy 12 detected by FISH were also detected by conventional cytogenetics.
- Risin et al. (1992) have karyotyped 36 lymphoblastoid cell lines established by Epstein-Barr virus (EBV) transformation. Of these, seven showed trisomy 12 including two lines with clonality for trisomy 12. As one explanation, the authors suggested that EBV transformation prone to the formation of trisomy 12 in certain individuals predisposed to development of lymphoproliferative diseases.
- Ceignet et al. (1993) applied FISH to the detection of trisomy 12 in 75 patients with CLL.
- An EBV-transformed lymphoblastoid cell line with a 48,XX,+X,+12 karyotype has been reported by Risin et al. (1993). The relevance of their findings with respect to trisomy 12 in CLL has been discussed.
- A patient initially diagnosed with refractory anemia with excess blasts in transformation underwent an evolution to a nonlymphocytic acute

leukemia. When the leukemia was diagnosed, bone marrow cells having a normal karyotype before showed a t(12;14)(q13;q32).

- The bone marrow cells of a patient with myelodysplastic syndrome-associated with eosinophilia showed a karyotype 45,XY,t(12;21)(q23;q22),−17. (Kobayashi et al. 1993).
- Herath et al. (1994) report on the cytogenetics of four hemangiopericytomas. They discuss anomalies of 7p15 and 12q as being important to the pathogenesis in cerebral (hemangiopericytomas).
- A case of an intracranial hemangiopericytoma with a recurrent t(12;19)(q13;q13.3) has been reported by Henn et al. (1993).

Mapping Data, Cell Hybrids

- Starting with a hybrid cell line containing the short arm of chromosome 12 as the only human material, Sinke et al. (1992) have employed an irradiation and fusion procedure to generate 16 somatic cell hybrids containing 12p-fragments. This panel encompasses the entire 12p.
- Baens et al. (1993) constructed an arrayed library of human heterogeneous nuclear complementary DNA from a sole 12p somatic cell hybrid.
- Mathew et al. (1992) have precisely mapped 13 single-copy DNA sequences previously assigned to the long arm of chromosome 12 by FISH.
  In addition, the results show that single copy DNA probes as small as 500 bp can be mapped by FISH.
- O'Brien et al. (1993) have provided data on comparative gene mapping between human chromosomes and other mammalian species.
- The results of the first international workshop on human chromosome 12 mapping (Oxford, 16–18 September 1992), particularly the progress made since HGM 11, are reported by Craig et al. (1992, 1993).
- A physical map of a 1.3 Mb region on the long arm of chromosome 12 spanning the GLI and LRP locus was constructed by Forus and Myklebost (1992) using pulsed-field gel electrophoresis.
- The gene for the human transcription factor protein Sp1 was mapped to 12q13 by FISH (Matera and Ward 1993)

References

Aman P, Ron D, Mandahl N, Fioretos T, Heim S, Arheden K, Willen H, Rydholm A, Mitelman F (1992) Rearrangement of the transcription factor gene CHOP in myxoid liposarcoma with t(12;16)(q13;p11). Genes Chrom Cancer 5:278–285

Atkin NB, Fox MF, Baker MC, Jackson Z (1993) Chromosome 12-containing markers, including two dicentrics, in three i(12p)-negative testicular germ-cell tumors. Genes Chrom Cancer 6:218–221

Badia L, Alvarez MA, Palau F, Prieto F (1993) Translocation (12;14)(q13;q32) in myelodysplastic syndrome. Cancer Genet Cytogenet 65:76–78

Baens M, Chaffanet M, Cassiman J-J, Van den Berghe H, Marynen P (1993) Construction and evaluation of a hncDNA library of human 12p transcribed sequences derived from a somatic cell hybrid. Genomics 16:214–218

Belge G, Kazmierczak B, Meyer-Bolte K, Bartnitzke S, Bullerdiek J (1992) Expression of SV40 T-antigen in lipoma cells with a chromosomal translocation t(3;12) is not sufficient for direct immortalization. Cell Biol Int Reports 16:339–347

Bullerdiek J, Wobst G, Meyer-Bolte K, Chilla R, Haubrich J, Thode B, Bartnitzke S (1993) Cytogenetic subtyping of 220 salivary gland pleomorphic adenomas: correlation to occurrence, histological subtype, and in vitro cellular behavior. Cancer Genet Cytogenet 65:27-31

Craig I, Gemmill R, Kucherlapati R (1992) Report of the first international workshop on human chromosome 12 mapping. Cytogenet Cell Genet 61:243–262.

Craig IW, Gedde-Dahl T, Gemmill R, Kucherlapati R (1993) Report of the committee on the genetic constitution of chromosome 12. Genome Priority Reports 1:402–418

Coignet L, Bertheas MF, Vasselon C, Jaubert J, Reynaud J, Calmard-Oriol P, Brizard CP, Guyotat D (1993) Trisomy 12 in B-cell chronic lymphocytic leukemia: interphase study by ISH in 75 patients. Nouv Rev Fr Hematol 35:129–134

Dal Cin P, Kools P, De Jonge I, Moerman P, Van de Ven W, Van den Berghe H (1993) Rearrangement of 12q14–15 in pulmonary chondroid hamartoma. Genes Chrom Cancer 8:131–133

Döhner H, Pohl S, Bulgay-Mörschel M, Stilgenbauer S, Bentz M, Lichter P (1993) Trisomy 12 in chronic lymphoid leukemias – a metaphase and interphase cytogenetic analysis. Leukemia 7:516–520

Edler von Eyben F, de Graaff WE, Marrink J, Blaabjerg O, Sleijfer DT, Schraffordt-Koops H, Oosterhuis JW, Petersen PH, van Echten-Arends J, de Jong B (1992) Serum lactate dehydrogenase isoenzyme activity in patients with testicular germ cell tumors correlates with the total number of copies of the short arm of chromosome 12 in the tumor. Mol Gen Genet 235:140–146

Fletcher JA (1992) Translocation (12;22)(q13–14;q12) is a nonrandom aberration in soft-tissue clear-cell sarcoma. Genes Chrom Cancer 5:184

Forus A, Myklebost O (1992) A physical map of a 1.3-Mb region on the long arm of chromosome 12, spanning the GLI and LRP Loci. Genomics 14:117–120

Gemmill RM, Mendez MJ, Dougherty CM, Paulien S, Liao M, Mitchell D, Jankowski SA, Trent JM, Berger C, Sandberg AA, Meltzer PS (1992) Isolation of a yeast artifical chromosome clone that spans the t(12;16) translocation breakpoint characteristic of myxoid liposarcoma. Cancer Genet Cytogenet 62:166–170

Henn W, Wullich B, Thonnes M, Stendel W-I, Feiden W, Zang K (1993) Recurrent +(12:19)(q13;q133) in intracranial and extracranial hemangiopericytoma. Cancer Genet Cytogenet 71:151–154

Herath SE, Stalboerger PG, Dahl RJ, Parisi JE, Jenkins RB (1994) Cytogenetic studies of four hemangiopericytomas. Cancer Genet Cytogenet 72:137–140

Hirabayashi Y, Yoshida MA, Ikeuchi T, Ishida T, Kojima T, Higaki S, Machinami R, Tonomura A (1992) Chromosome rearrangements at 12q13 in two cases of chondrosarcomas. Cancer Genet Cytogenet 60:35–40

Hu J, Surti U, Tobon H (1992) Cytogenetic analysis of an uterine lipoleiomyoma. Cancer Genet Cytogenet 62:200–202

Kashimura M, Minamihisamatsu M (1993) Chromosomal rearrangement, t(12;22)(p13;q13), in acute myelomegakaryoblastic leukemia. Cancer Genet Cytogenet 65:81–82

Kato A, Kamiyama T, Yamaguchi S, Sudo K, Mori S, Sekiguchi M, Aoki N, Shibuya M (1993) Establishment of a human small cell lung carcinoma cell line carrying amplification of c-myc gene and chromosomal translocation of t(3p;6p) and t(12q;17p). Jpn J Cancer Res 84:355–359

Kazmierczak B, Thode B, Bartnitzke S, Bullerdiek J, Schloot W (1992) Pleomorphic adenoma cells vary in their susceptibility to SV40 transformation depending on the initial karyotype. Genes Chrom Cancer 5:35–39
Kobayashi H, Kitano K, Shimodaira S, Ishida F, Saito H (1993) Eosinophilia in myelodysplastic syndrome with a (12;21)(q23;q22) translocation. Cancer Genet Cytogenet 68:95–98
Kools PFJ, Wanschura S, Schoenmakers EFPM, Yeurts JWM, Mols R, Kazmierczak B, Bullerdiek J, Van den Berghe H, Van de Van WJM (1994) Identification of yeast artificial chromosomes spanning the chromosome 12 translocation breakpoint of a pleomorphic salivary gland adenoma with t(1;12)(p22;q15). Cancer Genet Cytogenet (in press)
Krance RA, Raimondi SC, Dubowy R, Estrada J, Borowitz M, Behm F, Land VJ, Pullen J, Carroll AJ (1992) t(12;17)(p13;q21) in early pre-B acute lymphoid leukemia. Leukemia 6:251–255
Looijenga LHJ, Gillis JM, Van Putten WLJ, Oosterhuis JW (1993) In situ numeric analysis of centromeric regions of chromosome 1, 12 and 15 of seminomas, nonseminomatous germ-cell tumors and carcinoma in situ of human testis. Lab Invest 68:211–219
Mahmoud H, Carroll AJ, Behm F, Raimondi SC, Schuster J, Borowitz B, Land V, Pullen DJ, Vietti TJ, Christ W (1992) The nonrandom dic(9;12) translocation in acute lymphoblastic leukemia is associated with B-progenitor phenotype and an excellent prognosis. Leukemia 6:703–707
Mandahl N, Willen H, Rydholm A, Heim S, Mitelman F (1993) Rearrangement of band q13 on both chromosomes 12 in a periosteal chondroma. Genes Chrom Cancer 6:121–123
Mandahl N, Örndal C, Heim S, Willen H, Rydholm A, Bauer HCF, Mitelman F (1993) Aberrations of chromosome segment 12q13–15 characterize a subgroup of hemangiopericytomas. Cancer 71:3009–3013
Matera AG, Ward DC (1993) Localization of the human Sp1 transcription factor gene to 12q13 by FISH. Genomics 17:793–794
Mathew S, Murty VVVS, Hunziker W, Chaganti RSK (1992) Subregional mapping of 13 single-copy genes on the long arm of chromosome 12 by FISH. Genomics 14:775–779
Meloni AM, Surti U, Contento AM, Davare J, Sandberg AA (1992) Uterine leiomyomas: Cytogenetic and histologic profile. Obstet Gynecol 80:209–217
Mitchell ELD, Kelsey A, Armstrong G, Dickson A (1992): Ring chromosome in a retroperitoneal lipoma of childhood. Genes Chrom Cancer 4:89–91
Mrozek K, Karakousis CP, Perez-Mesa C, Bloomfield CD (1993) Translocation t(12;22) (q13;q12.2–12.3) in a clear-cell sarcoma of tendons and aponeuroses. Genes Chrom Cancer 6:249–252
Mrozek K, Karakousis CP, Bloomfield CD (1993) Chromosome 12 breakpoints are cytogenetically different in benign and malignant lipogenic tumors: localization of breakpoints in lipoma to 12q15 and in myxoid liposarcoma to 12q13.3. Cancer Res 53:1670–1675
Murty VVVS, Houldsworth J, Baldwin S, Reuter V, Hunziker W, Besmer P, Bosl G, Chaganti RSK (1992) Allelic deletions in the long arm of chromosome 12 identify sites of candidate tumor suppressor genes in male germ-cell tumors. Procl Natl Acad Sci USA 89: 11006–11010
O'Brien SJ, Peters J, Searle A, Womack J, Marshall-Graves J (1993) Report of the committee on comparative gene mapping. Genome Priority Reports 1:140–146
Ohjimi Y, Iwasaki H, Kaneko Y, Ishiguro M, Ohgami A, Fujita C, Shinohara N, Yoshitake K, Kikuchi M (1992) Chromosome abnormalities in liposarcomas. Cancer Genet Cytogenet 65:81–82
Qumsiyeh MB, Tharapel SA (1992) Interphase detection of trisomy 12 in B-cell chronic lymphocytic leukemia by fluorescence hybridization in situ. Leukemia 6:602–605
Raghoebier S, Kibbelaar RE, Keiverda JK, Klui-Nelemans JC, Van Krieken JHJM, Kokj F, Kluin PM (1992) Mosaicism of trisomy 12 in chronic lymphocytic leukemia detected by nonradioactive ISH. Leukemia 6:1220–1226

Reilly PA, Heerema NA, Sledge GW, Palmer CG (1993) Unusual distribution of chromosome 12 in a testicular germ-cell tumor cell line (833K) and its cisplatin-resistant derivative (64CP9). Cancer Genet Cytogenet 68:114–121

Risin S, Hopwood VL, Pathak S (1992) Trisomy 12 in Epstein-Barr Virus-transformed lymphoblastoid cell lines of normal individuals and patients with nonhematologic malignancies. Cancer Genet Cytogenet 60:164–169

Risin S, Fujimaki T, Mestriner CA, Brown CA, Brown NM, Hopwood VL, Fidler LJ, Pathak S (1993) Clonal expansion of cells with trisomy of chromosome 12 and X in an EBV-transformed lymphoblastoid cell line and establishment of a tumorigenic monoclonal cell line (48,XX,+X,+12). Cytogenet Cell Genet 62:54–55

Rodriguez E, Mathew S, Mukherjee AB, Reuter VE, Bosl GJ, Chaganti RSK (1992) Analysis of chromosome 12 aneuploidy in interphase cells from human male germ-cell tumors by FISH. Genes Chrom Cancer 5:21–29

Rodriguez E, Sreekantaiah C, Reuter VE, Motzer RJ, Chaganti RSK (1992) t(12;22)(q13;q13) and trisomy 8 are nonrandom aberrations in clear-cell sarcoma. Cancer Genet Cytogenet 64:107–110

Rohen C, Bonk U, Staats B, Bartnitzke S, Bullerdiek J (1993) Two human breast tumors with translocations involving 12q13–15 as the sole cytogenetic abnormality. Cancer Genet Cytogenet 69:69–71

Schoenmakers EFPM, Kools PFJ, Kazmierczak B, Bullerdiek J, Claussen U, Horsthemke B, Van den Berghe H, Van de Ven WJM (1993) Isolation of a somatic cell hybrid retaining the der(16)t(12:16)(q13;p11.2) from a myxoid liposarcoma cell line. Cytogenet Cell Genet 63:8–16

Schoenmakers EFPM, Kools PFJ, Mols R, Kazmierczak B, Bartnitzke S, Bullerdiek J, Dal Cin P, De Jong PJ, Van den Berghe H, Van de Ven WJM (1994) Physical mapping of chromosome 12q breakpoints in lipoma, pleomarphic salivary gland adenoma, uterine leiomyoma, and myxoid liposarcoma. Genomics 20:210–222

Schoenmakers EFPM, Mols R, Wanschura S, Kools PFJ, Yeurts JMW, Bartnitzke S, Bullerdiek J, Van den Berghe H, Van de Ven WJM (1994) Identification, molecular cloning and characterization of the chromosome 12 breakpoint cluster region of uterine leiomyomas. Genes Chromosomes Cancer (in press)

Schofield DE, Fletcher JA (1992) Trisomy 12 in pediatric granulosa-stromal cell tumors. Am J Pathol 141:1265–1269

Sinke RJ, Suijkerbuik RF, Herbergs J, Janssen H, Cassiman JJ, Van Kessel AG (1992) Generation of a panel of somatic cell hybrids containing fragments of human chromosome 12p by X-ray irradiation and cell fusion. Genomics 12:206–213

Sinke RJ, Suijkerbuik RF, De Jong B, Oosterhuis JW, Van Kessel AG (1993) Uniparental orign of i(12p) in human germ-cell tumors. Genes Chrom Cancer 6:161–165

Speleman F, Laureys G, Benoit Y, Cuvelier C, Suikerbuijk R, De-Jong B (1992) i(12p) in a near-diploid mature ovarian teratoma. Cancer Genet Cytogenet 60:216–218

Sreekantaiah C, Rao UNM, Sandberg AA (1992) Complex karyotypic aberrations, including i(12p), in malignant mixed Mullerian tumor of uterus. Cancer Genet Cytogenet 60:78–81

Stephenson CF, Berger CS, Leong SPL, Davis JR, Sandberg AA (1992) Analysis of a giant marker chromosome in a well-differentiated liposarcoma using cytogenetics and FISH. Cancer Genet Cytogenet 61:134–138

Stern C, Deichert U, Thode B, Bartnitzke S, Bullerdiek J (1992) Eine zytogenetische Subtypisierung von 139 Uterus-Leiomyomen. Geburtsh Frauenheilk 52:767–772

Suijkerbuijk RF, Looijenga L, De Jong B, Oosterhuis JW, Cassiman JJ, Guerts Van Kessel A (1992) Verification of isochromosome 12p and identification of other chromosome 12 aberrations in gonadal and extragonadal human germ-cell tumors by bicolor double FISH. Cancer Genet Cytogenet 63:8–16

Tayeb MT, Neff JR, Bridge JA (1993) A case report of fibrolipoma with t(12;16)(q13;q24). Cancer Genet Cytogenet 67:145–146

Travis JA, Bridge JA (1992) Significance of both numerical and structural chromosomal abnormalities in clear-cell sarcoma. Cancer Genet Cytogenet 64:104–106
Wessels JW, Fibbe WE, Van der Keur D, Landegent JE, Van der Plas DC, Ottolander GJ, Roozendaal KJ, Beverstock GC (1993) t(5;12)(q13;p12) a clinical entity with features of both myeloid leukemia and chronic myelomonocytic leukemia. Cancer Genet Cytogenet 65:7–11

Subject Index

MIX
Papier aus verantwortungsvollen Quellen
Paper from responsible sources
FSC® C105338

If you have any concerns about our products,
you can contact us on
ProductSafety@springernature.com

In case Publisher is established outside the EU,
the EU authorized representative is:
Springer Nature Customer Service Center GmbH
Europaplatz 3, 69115 Heidelberg, Germany

Printed by Libri Plureos GmbH
in Hamburg, Germany